AF618948

Smart Sensors, Measurement and Instrumentation

Volume 24

Series editor

Subhas Chandra Mukhopadhyay
Department of Engineering, Faculty of Science and Engineering
Macquarie University
Sydney, NSW
Australia
e-mail: S.C.Mukhopadhyay@massey.ac.nz

More information about this series at http://www.springer.com/series/10617

Hemant Ghayvat · Subhas Chandra Mukhopadhyay

Wellness Protocol for Smart Homes

An Integrated Framework for Ambient Assisted Living

Hemant Ghayvat
Massey University
Palmerston North
New Zealand

Subhas Chandra Mukhopadhyay
Department of Engineering
Faculty of Science and Engineering
Macquarie University
Sydney, NSW
Australia

ISSN 2194-8402 ISSN 2194-8410 (electronic)
Smart Sensors, Measurement and Instrumentation
ISBN 978-3-319-52047-6 ISBN 978-3-319-52048-3 (eBook)
DOI 10.1007/978-3-319-52048-3

Library of Congress Control Number: 2016962039

Printed on acid-free paper

This Springer imprint is published by Springer Nature
The registered company is Springer International Publishing AG
The registered company address is: Gewerbestrasse 11, 6330 Cham, Switzerland

To our parents:
Late. Sree Bhaskar Ghayvat and
Mrs. Sadhana Ghayvat,
Mr. D.N. Mukhopadhyay
and Mrs. R.R. Mukhopadhyay

To my wife (Hemant): Arpita Hemant Ghayvat

To my wife (Subhas): Dr. K.P. Jayasundera

To my children (Subhas):
Miss. Sakura J. Mukhopadhyay
and Master Hiroshi J. Mukhopadhyay

To my brother (Hemant): Basant Ghayvat

Preface

This book covers the design and development of a Wellness Protocol that forecasts the wellness of individuals living in AAL environment. The Protocol is based on wireless sensors and networks that are applied to data mining and machine learning to monitor the activities of daily living. The heterogeneous sensor and actuator nodes, based on WSNs, are deployed in the home environment. These nodes generate the real-time data on object usage and other movements inside the home, in order to forecast the wellness of any individual. The new Protocol has been designed and developed to be suitable especially for smart home systems. The protocol has proven to offer a reliable, efficient, flexible, and economical solution for smart home systems and AAL environment.

According to consumers' demand, the Wellness Protocol-based smart home systems can be easily installed with existing households without any significant changes and with a user-friendly interface. Additionally, the Wellness Protocol has led to designing a smart building environment for apartments. In the endeavour of smart home design and implementation, the Wellness Protocol deals with large data handling and interference mitigation. A Wellness Protocol-based smart home monitoring system is the application of automation with integral systems of accommodation facilities to boost and enhance the everyday life of an occupant.

Palmerston North, New Zealand — Hemant Ghayvat
Sydney, Australia — Subhas Chandra Mukhopadhyay

Contents

Chapter 1
Introduction

Abstract In a smart home living environment, technology assists the occupants in their daily life. With the introduction of sensors, embedded processors and wireless communication technology, normal homes are converted into smart homes.

1.1 Introduction

In a smart home living environment, technology assists the occupants in their daily life. With the introduction of sensors, embedded processors and wireless communication technology, normal homes are converted into smart homes. In a smart home, the activities of an occupant are monitored continuously. Help or assistance is provided at the time of need or in an unforeseen situation. The current chapter provides the motivation behind this research, problem formulation, solution and the novel contribution of the research.

1.2 Motivation of Designing Homes for Tomorrow

The present research of design and implementation of An Integrated Framework for Smart Homes has the motivation to support the independent living of an occupant with utmost comfort, care, and safety.

1.2.1 Independent Living

The number of people living alone at home has been steadily increasing in the recent past and will continue in the near future. Moreover, it is a phenomenon that is spread across the globe (DePaula 2014). Figures and statistics do not explain the full story, although according to the research firm Euromonitor International, the

H. Ghayvat and S.C. Mukhopadhyay, *Wellness Protocol for Smart Homes*,
Smart Sensors, Measurement and Instrumentation 24,
DOI 10.1007/978-3-319-52048-3_1

Table 1.1 The statistic of householders living alone; it includes all age groups (Statistics Brain 2015)

Living alone numbers (as a percent of all households)		
Country		Percent of households (%)
1	Sweden	47
2	Britain	34
3	Japan	31
4	Italy	29
5	U.S.	28
6	Canada	27
7	Russia	25
8	South Africa	24
9	Kenya	15
10	Brazil	10
11	India	3

Source Euromonitor International, U.S. Census

number of people living alone all around the world is climbing sharply (Statistics Brain 2015). Table 1.1 represents the humans living alone in many parts of the world.

Sweden has more solo living than any other country in the world, with 47%, followed by the UK with 34%. The most surprising figure in the list comes from Japan with 31%, because Japan is well known for its organized family structure. Euromonitor International forecasts that the world will add up 48 million new single households by 2020, a sharp increase of 20%, as shown in Fig. 1.1 (Baker 2012).

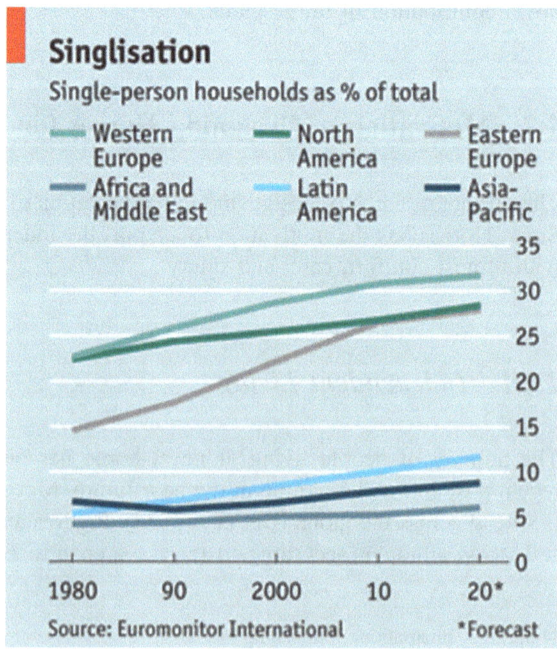

Fig. 1.1 Single-person households (Baker 2012)

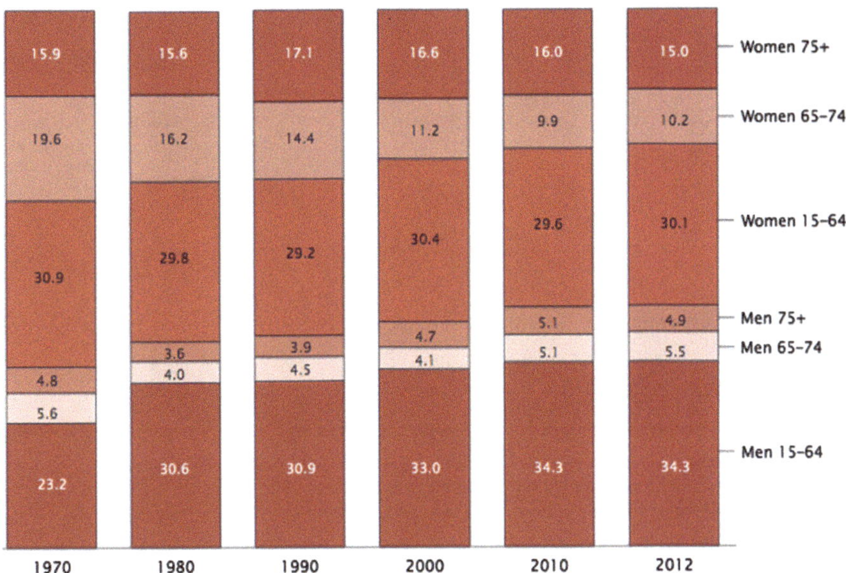

Fig. 1.2 One person households by Age and Sex, 1970 to 2012 in the US (Badger 2013)

It is believed that elderly people live alone at home. From the records of a US firm, it is seen that the percentage of one-person household aged between 15 and 64 is higher than the 65 plus, as shown in Fig. 1.2; while Fig. 1.3 represents the trend of single living households (Badger 2013).

Many heart-rending incidents have been reported about people living alone. A 37-year-old woman, Kirstine Hill, lay dead in her Hastings home for up to three weeks before her death was revealed [*The New Zealand Herald, March 2013*]. Police found her body in the kitchen of her semi-detached flat. A neighbor said afterward: "As most people live here, we keep to ourselves" (Ellingham and Raethel 2013). An elderly man living alone was rescued after spending days lying injured on his kitchen floor at Palmerston North, New Zealand (King 2015) in May, 2015. Such heart-breaking incidents do happen, and many are not recorded.

Technology assisted homes may be able to prevent such incidents, or at least the news of abnormal incidents will be reported quickly.

1.2.2 Enhance the Comfort and Lifetime

In the 21st century, scientific developments have allowed us to enhance the home environment. One of the key intentions of smart home study is to ease everyday life

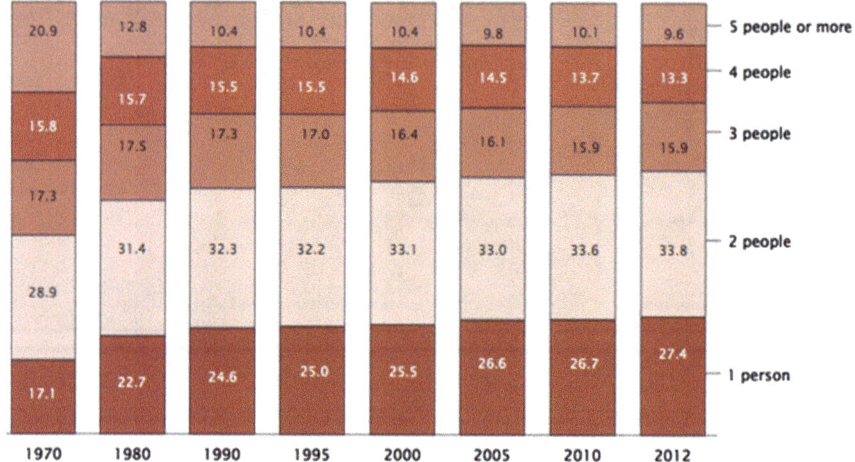

Fig. 1.3 Trend of living alone in the US (Badger 2013)

by increasing user comfort. Comforts achieved in two ways: One is associated with human activity identification. Understanding the user behavior in the home ambient is useful in making the home environment smarter and more relevant to needs. The other is remote home monitoring and control (Marmot and Bell 2012).

1.2.3 Health Services

Intelligent ambient integrated homes are already supporting the health care. In the case of any health care emergency, they provide the support through alerts to caregivers or healthcare professionals (Häfner et al. 2012). Moreover, they are monitoring people who are recovering from sickness. Smart home assisted living allows an occupant to manage an independent lifestyle (Kinsella et al. 2013).

1.2.4 Efficient Use of Electricity

One of the applications of smart home technology is to enhance the people's comfort by maximizing power savings (Raworth 2012). Comprehensive organization and information about an individual's activities in daily life play a significant role in achieving this goal. Lifestyle patterns help to develop flexible, dynamic prognostic, and conflict resolution proficiencies to the smart home automation system (Novitzky 2015).

1.2.5 Safety and Security

Safety and security are crucial traits of human life. Hence, combining safety measures in the smart home is a vital necessity for most of the smart home occupants. Typical homes are exposed to security threats. Most of the security issues are related to pathetic user access as well as device authentication arrangements. These are attacks either generated locally, such as robbery, or virtually, by accessing the smart home data (Bisio 2015).

1.3 Problem Formulation

Although the smart home technology is growing fast and advancing, it has to address a few significant issues such as packet reliability, device failures, and delay in decision-making information generation. Occupants demand that the smart home system should be reliable, efficient and user-friendly.

Heterogeneous sensing units design smart home systems and the data from these sensors are collected into local home gateway server. These local home gateways are not the high-speed computer, which can store large data and process them in a fraction of a second. Every sensor sends the data after a defined sampling time. It means these data packets are independent of event occurrence and produce excess data. It is not feasible for an occupant to maintain the system due to large data. In the smart home sensor, the node sends data according to their characteristics. The processing of this excess sensing data signal generates contention and leads to delay. There must be data priority for an emergency data signal. For instance, during a panic button press, the care support signal must reach immediately at support specialists without any delay. In these kinds of cases, priority and event-based approaches are inevitable features to resolve them.

The home environment has many unwanted sources of electromagnetic radiation that degrade the reliability and performance of the WSN-based smart home. There are intra-network interference and multi-path losses that affect packet delivery. Electromagnetic interference is the disruption that upsets desired node signal processing. This intra-network interference turns out to be more critical in unregulated free ISM band of the frequency spectrum.

By the influence of Internet of the Things (IOT), the daily usages objects are uniquely addressable and interconnected. The IOT enabled home monitoring has allowed connecting the information to the world wide web. The Home monitoring system demands user authentication to prevent unauthorized access (Wang et al. 2015; Butala et al. 2012; Stauffer 1991).

The brain of the smart home is data mining and machine learning algorithms which extract the useful information from heterogeneous sensing data for decision making and assistance to an occupant. There are a number of algorithms such as the HMM, CRF, and Naïve Bayes models, but their trend and forecasting for realistic uncontrolled home environments are not always satisfactory (Wacks 1993; Ricquebourg et al. 2006). Some of these models are either too simple and fail to cover all possibilities for pattern generation and forecast, or too complex and consume significant processing time. The smart home algorithm needs to be flexible and adaptive enough for seasonal and sudden activities change. In the smart home system, any anomalous behavior is detected due to any of the following two reasons: firstly, significant distinctions in the behavior of the occupant and secondly systems behavior change over the time due to equipment failure. Diagnosis of anomalous behavior is inevitable to avoid fake alerts.

According to the above discussion the key challenges identified in this research, are:

- Diverse hardware and software based on different technologies and standards cause issues in the process of integration.
- The immediate requirement of a customized smart home protocol to offer priority and event-based model.
- A large data handling issue arises in the heterogeneous wireless sensing system.
- The packet reliability parameter degrades, due to interference, congestion, equipment failure and multi-paths. This degradation causes an adverse effect on smart home system performance.
- Processing time becomes crucial in the case of smart home safety, security and healthcare applications.
- Maintaining the privacy of inhabitant's data and making authentication for caregiver only is highly recommended.
- Diagnosing the behavior and detecting the abnormality with the least possible fake alerts.

1.4 Problem Solution

To address the issues, a smart home needs a new protocol and standard which is customized to offer AAL and solutions. Every sensing system needs some basic modules; one of them is the RF protocol for wireless communication. The wellness system, therefore, started with ZigBee. AAL with ZigBee protocol was implemented in a realistic home condition, and the results were encouraging, albeit, there were some limitations of ZigBee protocol that we observed and analyzed.

These limitations were associated with the hardware and software fundamentals of ZigBee, which made it excellent for environmental monitoring but bulky for smart home requirements. The problem was associated with the ZigBee protocol structure. ZigBee is well known for environmental monitoring of a relatively large area, but it is not customized for the smart home. Some of the prominent issues that adjudicate the performance in the monitoring system were recognized during the wellness pattern generation, and forecasting with the ZigBee-based smart home system as follows:

The ZigBee protocol, used in the past, was a generalized protocol. The protocol provided the possibility of adding a huge number of nodes (65,535) and forced the use of many data for device identification. The routing protocol was designed by the manufacturer (proprietary protocol) and did not allow us to include anything of our own (Ricquebourg et al. 2006; Ghayvat et al. 2016).

- The ZigBee was not able to perform any customized and intelligent data process. It was limited up to the radio communication.
- The header data were too large, which included many fields according to the big area, but were not required in a moderate home space.
- It did not contain any intelligence and control for data processing. Vast amounts of data were transmitted, creating a problem.
- The sampling and storage were also a significant issue.

These deficiencies of previous research and the challenges offered by customer's demands have led us to build a modular, service-oriented new protocol, specially designed to support the development of pervasive computing spaces. The new communication protocol has been designed for the AAL. The Wellness protocol is an approach which targets an event and a priority-based communication. This protocol offers a complete smart home solution, starting from the sensor node to real-time analysis, data streaming, decision-making, and control.

The AAL requires activity recognition of daily living from raw sensor data, and these raw sensing data sets are composite and asymmetrical to encode into predefined scenarios. Even after encoding this raw data, it was somewhat difficult to recognize deviant behavior since these sensing data sets are on different time measures (sampling rate) and sense modalities. These diverse time and sensor modalities generate complication sin the course of equation formulation. It is rather easy to generate behavioral patterns from single event data sets. However, it may increase false alerts. The integrated framework based wellness time series function analyzed the individual's behavior from the previous data, real-time recently received data, and feedback received data. It reduces the false alarm to a significant extent. Thus, pattern detection through pipeline processing with filtering, characteristic construction, activity detection, and smoothing was done. An attempt was made to separate the routine data from unexpected data, which may have put an occupant's health and wealth at risk.

Moreover, the concerns of interference and deployment were adequately investigated, and mitigation approaches were developed to deal with the interference sources in the home environment.

1.5 Reasoning of Wellness Protocol and Approach

- The wellness protocol-based integrated framework for AAL is to monitor the behavior of an occupant in real time.

- It assists the occupants with their independent living and obtains help in case of any emergency or panic. Moreover, it makes it easier for distant caregivers, worried about the occupant.
- The communication module of the wellness protocol is event and priority based, which sends the data to the server according to its significance and priority.
- The protocol provides the transmission control at node end through the intelligent sampling algorithm.
- The developed framework is accurate for sensing data storage and analysis in a real-time sensing system.
- The wellness determination and forecasting is based on wellness indices and belief time series methods.

1.6 Scope of Book

This book focuses mainly on the technical characteristics of smart homes.As smart home research in general spans multiple disciplines this emphasis will naturally limit the scope of the study to hardware, software, and communication. Research findings resulting from user trials and other issues interrelated to the activity of daily livings (ADLs)are also presented. Serviceability concerns and embedded design (e.g. PC software, signal processing algorithms) are also included as necessary. Smart ambient or indoor environment are more general terms for this kind of research, but in the perspective of this book the environment in question is a home as a physical space. Thus, this book is the research finding for Ambient Assisted Living. Outdoor environment, wearable computing (e.g. smart clothing) and other related technologies are beyond the scope of this book.

1.7 Novel Contribution

The novel contributions of this research are as follows:

- Integration of Advanced Wireless Sensing and Information Technology for the wellness evaluation of an occupant living in an assisted environment.
- Accurate wellness determination in real time as well as near real-time from streaming sensing data with the application of a time series model.
- Improved Wellness Function definitions (β_1 and β_2) based on Wellness Belief and Wellness Indices to monitor the activity of daily living.
- The integrated framework provides for control and monitoring of a smart home through the Internet of Things.
- Optimum packet reliability and system performance through an interference mitigation approach.

- An identification of ADLs through object usage and movement in the home environment for wellness pattern generation and forecasting.
- The integrated framework of the Wellness Protocol having functionalities to monitor the occupant in real time, and the electricity consumption indices of different household objects.
- Large data handling through an Intelligent Sampling Algorithm to reduce the computational time and resources.
- Event and Priority based packet encapsulation are designed to deal with various activities and conditions in the home environment.

1.8 Book Overview

This book contains the following seven chapters.

- Chapter 1—Introduction

Introduces the book, provides a summary of motivations, problem formulation, and solution, the scope of research and reasoning and novel contributions.

- Chapter 2—Literature Survey

Introduces the background to a smart home communication protocol, data mining, and machine learning approaches for activity recognition and forecasting and describes the user studies performed within this research.

- Chapter 3—Wellness Protocol Development and Implementation

The design and implementation details are described, providing an explanation of how the Wellness protocol was designed and functions. The details of sensing node development, device configuration, and deployment, wireless data communication, storage, and analysis approach of heterogeneous sensor data fusion have been documented.

- Chapter 4—Issues and Mitigation of WSNs Based Smart Building System

This research investigates the issues of interference based on wireless system deployment in a home environment and suggests the mitigation assist the smart building environment system designer in evaluating and measuring the on-site performance.

- Chapter 5—Activity detection and wellness pattern generation

This chapter includes the classification of activities, development of the Wellness belief, wellness function model and methodology to wellness pattern generation. In the end, the web-based results of the wellness system are shown.

- Chapter 6—Wellness Pattern Forecasting

This chapter describes the Wellness Pattern Forecasting for the occupants for upcoming days through the past datasets with the minimum false alerts.

- Chapter 7—Conclusion and Future Work

The conclusion of present research work and issues for future research work.

References

B. Depaula, (2014, June 23). *Where in the World Do People Live Alone?*. Retrieved Nov 18, 2016, from http://blogs.psychcentral.com/single-at-heart/2014/06/where-in-the-world-do-people-live-alone/ .

Statistics Brain, (2015, October 3). *Living Alone Statistics*. Retrieved Nov 10, 2016, from http://www.statisticbrain.com/living-alone-statistics/.

A. Baker, (2012, August 25). *The attraction of solitude*. Retrieved Nov 9, 2016, from http://www.economist.com/node/21560844.

E. Badger, (2013, August 27). *A Brief History of How Living Alone Came to Seem Totally Normal*. Retrieved Nov 10, 2016, from http://www.citylab.com/housing/2013/08/brief-history-how-living-alone-came-seem-totally-normal/6689/.

J. Ellingham and J. Raethel, (2013, July 7). *How does someone die alone. Retrieved* Nov 12, 2016, from http://www.stuff.co.nz/dominion-post/news/8888382/How-does-someone-die-alone.

K. King, (2015, May 28), *Elderly man rescued after spending days lying injured on his kitchen floor*. Retrieved Nov 12, 2016, from http://www.stuff.co.nz/manawatu-standard/news/68869717/elderly-man-rescued-after-spending-days-lying-injured-on-his-kitchen-floor.

M. Marmot and R. Bell, "Fair society, healthy lives," *Public Health,* vol. 126, pp. S4-S10, 2012.

S. Häfner, J. Baumert, R. Emeny, M. Lacruz, M. Bidlingmaier, M. Reincke, *et al.*, "To live alone and to be depressed, an alarming combination for the renin–angiotensin–aldosterone-system (RAAS)," *Psychoneuroendocrinology,* vol. 37, pp. 230–237, 2012.

K. Kinsella, J. Beard, and R. Suzman, "Can Populations Age Better, Not Just Live Longer?," *Generations,* vol. 37, pp. 19–26, 2013.

K. Raworth, "A safe and just space for humanity: can we live within the doughnut," *Oxfam Policy and Practice: Climate Change and Resilience,* vol. 8, pp. 1–26, 2012.

P. Novitzky, A. F. Smeaton, C. Chen, K. Irving, T. Jacquemard, F. O'Brolcháin, *et al.*, "A review of contemporary work on the ethics of ambient assisted living technologies for people with dementia," *Science and engineering ethics,* vol. 21, pp. 707–765, 2015.

I. Bisio, F. Lavagetto, M. Marchese, and A. Sciarrone, "Smartphone-centric ambient assisted living platform for patients suffering from co-morbidities monitoring," *Communications Magazine, IEEE,* vol. 53, pp. 34–41, 2015.

K. Wang, Y. Shao, L. Shu, G. Han, and C. Zhu, "LDPA: a local data processing architecture in ambient assisted living communications," *Communications Magazine, IEEE,* vol. 53, pp. 56–63, 2015.

P. M. Butala, Y. Zhang, T. D. Little, and R. C. Wagenaar, "Wireless system for monitoring and real-time classification of functional activity," in *Communication Systems and Networks (COMSNETS), 2012 Fourth International Conference on*, Bangalore India, 2012, pp. 1–5.

H. B. Stauffer, "Smart enabling system for home automation. Consumer Electronics," *IEEE Transactions on,* vol. 37, pp. xxix-xxxv, 1991.

P. Wacks, "The impact of home automation on power electronics," in *Applied Power Electronics Conference and Exposition, 1993. APEC'93. Conference Proceedings 1993., Eighth Annual,* San Diego, CA, 07–11 Mar 1993, pp. 3-9.

V. Ricquebourg, D. Menga, D. Durand, B. Marhic, L. Delahoche, and C. Loge, "The smart home concept: our immediate future," in *E-Learning in Industrial Electronics, 2006 1ST IEEE International Conference on*, Hammamet, 18–20 Dec, 2006, pp. 23-28.

H. Ghayvat, S. Mukhopadhyay, and X. Gui, "Issues and mitigation of interference, attenuation and direction of arrival in IEEE 802.15. 4/ZigBee to wireless sensors and networks based smart building," *Measurement,* vol. 86, pp. 209–226, 2016.

Chapter 2
Literature Survey

Abstract The present chapter presents the most relevant research approach and methodologies based on WSNs-based home monitoring system for Ambient Assisted Living (AAL). There are some research works, but this section includes only the researchers who give full understanding and justify the work.

2.1 Introduction

The selection criteria for this review are shown in Table 2.1. The chapter has been included if they are published in English, in a peer-reviewed text, and are available as full works. Because of the rapid progression in technology and the relative lack of information in earlier years, articles published before January 2005 have been excluded. This research has focused on novel evidence regarding the success or feasibility of smart-home technologies. Consequently, the search has been limited to intervention or practicality studies. Chronicle evaluations and other methodical reviews have been excluded as they do not fulfill inclusion criteria. For the purpose of this literature survey, studies have been considered evaluating the efficiency of the smart home technique when they have combined an intrusion period with evaluation before and afterward. A 'home' environment is the place where a person lives; this may be a private home, supported public home, assisted living and retirement house. Most of the people are not willing to do permanent changes in their living space, so, in few research the researchers have used either laboratory setup which is purpose built home. Such settings have also been included as people are able to live in the 'house,' and the setting has been consequently considered being a 'home' environment. Hospital environments and psychological assistance usually offer the assisted physiological environment to people who are under observation for recovery. Subsequently, studies in nursing homes and hospitals have been excluded. Throughout this study, a broad definition of different age groups has been considered a younger adult (aged below 45), middle-aged (aged 45–64 years), aged (65–79 years) and aged (80+ years). Tele-rehabilitation and telehealth mostly involve communication with a distant health specialist and is

H. Ghayvat and S.C. Mukhopadhyay, *Wellness Protocol for Smart Homes*,
Smart Sensors, Measurement and Instrumentation 24,
DOI 10.1007/978-3-319-52048-3_2

Table 2.1 The criteria filter for selecting the research in literature survey

Enclosure criteria	Elimination criteria
• Assessed smart-home models • Available in English and full-text from peer review journals • Assessed usefulness or practicability • Implemented in a home environment • For ambient assisted living	• Published before January 2005 • Realized in other living environments such as treatment center or rehabilitation settings • Books, degree theses and abstracts from conference presentations • Descriptive reviews and other systematic reviews

Table 2.2 The keyword used on Google Scholar and other research article search engine to find research studies

Keyword	Synonyms
Smart-home	"Smart home" or "ubiquitous technology" or "ubiquitous home" or "electronic assistive technology" or "ambient assisted living" or "telecare social alert platform" or "social alarm" or "environmental control system" or "Home Automation" or "Machine Learning" or "Data Mining" or "Eldercare" or "automated home environment" or "WSNs Protocol" or "Smart Building" or "Mitigation"

therefore still subject to the medical system for support but not such as hospital care (Wacks 1993; Badger 2013; DePaula 2014; Wang et al. 2015). Consequently, studies of telerehabilitation or telemedicine-based management techniques have been included up to a limited extent. Multiple forms of technology may help to assist an inhabitant in the home environments. Robotics, gaming or social inclusion literature, are beyond the scope of this review; so, have been excluded. The focus of this research review is specifically on types of techniques that can be used in a home environment which either interact with or provide direct information to the user without the need for another individual (Kinsella et al. 2013; Marmot and Bell 2012). Table 2.2, presents the keywords used in search engines stand-alone or in combination to collect the research findings of the WSNs based smart home monitoring system.

The rest of the chapter is subdivided into different sections of smart home system design and most significant modules of AAL.

2.2 Smart Home for Wellness

The smart home is a customized application of computational technology for the home. Numbers of equivalent names are used for the smart home system, e.g., home monitoring, home automation, assistive living system, intelligent home and smart home. Stauffer (1991) provided an early definition of smart homes.

According to Stauffer, SMART HOUSE is a comprehensive system that delivers the general resources required for home automation in a multi-product and the multi-vendor environment. A system controller, house extensive wiring network, communication protocols, standard interfaces (outlet designs) for joining other products, and basic user controls are the elements of the smart home system. This definition is more towards the automation and control, but they do not consider the monitoring. A similar kind of definition we find in Wacks (1993), Home Automation is the term of a new trade that supports the next generation of consumer electronic machines. The significant value added by home automation is the integration of products to produce new applications.

Smart homes are not only limited up to just turn on and off the device but also monitor the internal environment and the activities that take place while the house is occupied. The outcome of these amendments to the technology is that a smart home can now monitor the happenings of the occupant. Moreover, individually operate devices on a set of predefined patterns or independently as the user requires. Vincent has included the intelligence with home automation (Ricquebourg et al. 2006).

Taking into consideration the recent development in smart home technology, a smart home can be explained as an implementation of pervasive computing and sensing. This smart home technology is capable of delivering context-aware, automated and assistive services to home monitoring and remote control. Ambient Assisted Living (AAL) is one of the smart home applications, which comprises of interoperable concepts, products, and services, which integrate new information and communication technologies (ICT) and home environments with the aim to improve and enhance the quality of life for people in all steps of the life cycle. AAL can at better be understood as an individual's requirement-based support systems for a vigorous and liberated life that caters for the different abilities of their users. The framework of AAL is principally concerned with the individual in his or her immediate ambient by offering user-friendly interfaces for all kinds of objects in the home (Leone et al. 2015; He and Zeadally 2015; Dias et al. 2015).

2.3 Entities of Smart Home Systems

Smart processing and control platform based smart home system has gained tremendous interest and consideration of the industry by the fact of self-management and intelligent, customized setting for a particular application. Sensor with RF communication chip without any external intelligent processing platform is also popular in wireless sensor and systems application for monitoring the home, but external processing unit and control give an extra edge over the traditional approach of monitoring. With the innovation in micro electro mechanical systems (MEMS), digital electronics and advancement in modern wireless communication technology the intelligent wireless sensor and systems (IWSNs) have got the potential to monitor physical world, processing and handling of the raw data, creating useful and essential information from that data, based on the observations and executing respective command. The microelectronic sensing circuit measures the generic ambient parameters such as temperature, pressure, humidity, movement, and link to the location of the sensor and transmogrify into an electric signal. In IWSNs, there is always processor and controller logic before transforming this nascent ambient condition values into a wireless signal through conditioning circuitry. The IWSNs have a vast and varied use, and it is becoming an indispensable and vital section of modern technological systems, for instance, smart home monitoring, ambient assisted living, building automation, weather forecasting, nuclear reactor, heath services, and defense radar.

According to the desired application and characteristics of sensor networks, it prescribes adequate discipline and regulation on the IWSNs formulation and plan. Deployment environment and location, self-healing, fault recognition and tolerance, portability, flexibility, robustness, network topology, design cost, power consumption, communication protocol are some of the design characteristics. By the fact of these challenges, there is tremendous, and extensive research is pursuing on IWSNs. In most of the short-range applications, the designer prefers ISM (industrial, scientific and medical band) frequency spectrum, which is free of cost, unlicensed and unregulated; 2.4 GHz spectrum is universal in use. Figure 2.1 represents a general purpose block diagram of smart home functioning (Seo et al. 2016). In the smart home system, each module has its role to fulfill the smart home service stack. The modules of intelligent wireless sensor and networks based smart home for assisted living is explained in Fig. 2.1 (Enck et al. 2014; Palumbo et al. 2014).

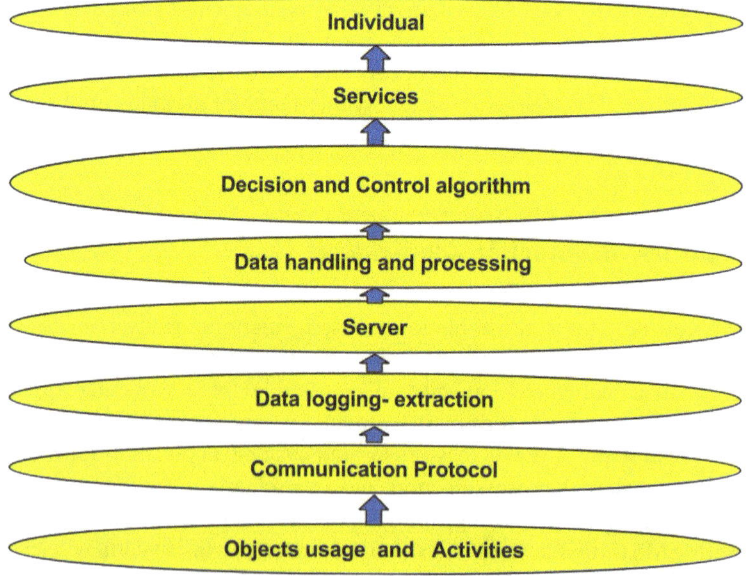

Fig. 2.1 General functioning block diagram of smart home system

2.3.1 Sensors and Actuators

A sensor node is a small, inexpensive device with reasonable sensing that transforms an ambient or biological measure into electrical form; this electrical signal is converted into digital form and forwarded to the microcontroller for remaining data processing at end-device side. While the actuator is a device to translate an electrical control signal to physical action and starts the event through which an agent acts upon the physical environment, by this way actuators are hardware entities. Moreover, an actuator being able to act on the physical world through one or several actuators is also a network unit that executes networking logics such as transmission, reception, and process data through this approach.

To satisfy the demand of a recent occupant, the smart home should be surrounded with a network of heterogeneous and multipurpose cost-efficient wireless sensing solution, able to sense the diverse range of parameters essential for the intelligent environment. Sensors and actuators can be classified into two groups as traditional and advanced types according to a recent development in the sensing world. Traditional sensors are temperature, pressure, humidity, force, light, motion and gas sensors. For instance, temperature and humidity sensors can be used in concurrence of air-conditioning either turning the fan, a heater as well as air-conditioning on/off. Light intensity sensor could be used to activate the light automatically at sunset, or darkness comes (Rodrigues et al. 2012). Advanced sensors are the sensors which are more than the ambient sensing and to look after the particular need of individuals such as emotion and impact sensor. For a context-aware scenario, it is necessary that it should identify the mood of the occupant, for instance, heart rate sensor discovers the pressure and generates medicine reminder as well as inform the caregiver as desired. To design the intelligent home environment sensing is essential, this means in a smart home different types of sensors need to be deployed. So, more the number of sensors more the area it covers and offers the best monitoring outcome. Some people prefer camera and RFID while others avoid this to reduce the cost (Dengler et al. 2007; Mukhopadhyay 2015).

Table 2.3 Ambient sensors to smart environment monitoring

Sensor	Measurement
PIR	Motion
Active infrared	Motion/Identification
RFID	Object information
Pressure	Pressure on mat, chair, etc.
Smart tiles	Pressure on floor
Magnetic switches/contact wire	Door/Cabinet opening/closing
Ultrasonic	Motion
Temperature	Environmental as well as subject temperature
Gas/Smoke	Gas or high temperature

There are different kinds of sensors available for home monitoring; Table 2.3 shows the ambient sensors applicable to Smart Environments.

2.3.2 Controller and Processing Unit

Each sensor produces an electrical signal, and this electrical signal is converted into digital form. The data generated by the sensor is raw information. This raw information needs to be processed further to perform certain essential actions. These actions are a direct function of the programming and configuration logic that implements in the processor. Different levels of intelligence can be designed and performed based on personalized requirement. The standard of intelligence can be significantly improved through the devices; devices interchange the decision information, and precision of this information sharing define the services in the smart home (Wang et al. 2013; Han et al. 2014).

2.3.3 Defined Wireless Networking Protocols

In an era of prompt advancement in digital technology, most of this technology is focused on proficient monitoring and controlling. Ubiquitously from mammoth structure building automation to a smart small home, big industrial assembly mechanisms to the tiny toy, an ordinary undergraduate laboratory to international space research center and even health care service at a desk through wireless sensor & networks, and WSN has become an indispensable and crucial device in playing an important role. The significant enhancement by introducing wireless technology is that it condenses and simplifies the medium complication and harness which comes in wired transmission; it offers facilitation for the installation of sensors, controllers, and actuators (Suryadevara et al. 2015). It provides generous benefits such as mobility, lower life-cycle cost and reducing connector failures, which is one of the most common reliability problems. The cost and installation labors with an increased number of sensors in an urban environment are exponentially decreasing by the wireless technology innovations. There are different wireless communication mediums (technology) in which wireless sensor networks can be constructed according to respective applications and strengths, such as WiFi; WiMax, Bluetooth, and ZigBee (Byun et al. 2012).

The IEEE Standards include a family of networking standards that cover the PHY and media access control (MAC) layer specifications for implementing WSN. IEEE 802.11 series regulate PHY and MAC layers for wide local area networks (WLAN) by employing radio frequency bands at 2.4 and 5.8 GHz. WiFi is a widespread IP-based wireless technique which is applied in the home, office, in consumer appliances such as a mobile phone. Furthermore, advanced security features and the QoS can also be extended using Wi-Fi technology. The IEEE

Table 2.4 Wireless network based on IEEE standard (Mendes et al. 2015)

Protocols/Specifications	ZigBee	WirelessHART	MiWi Over	Isa100.11	Bluetooth	Wi-Fi IEEE 802.11 a/b/g/n/ac/i
ISM bands	2.4 GHz, 915 MHz (USA), 868 MHz (EU)	2.4 GHz, 915 MHz (USA), 868 MHz (EU)	2.4 GHz, 915 MHz (USA), 868 MHz (EU)	2.4 GHz, 915 MHz (USA), 868 MHz (EU)	2.4 GHz	2.4 GHz, 5 GHz
Number of radio channels	16 (2.4 GHz), 10 (915 MHz), 1 (868 MHz)	16 (2.4 GHz), 10 (915 MHz), 1 (868 MHz)	16 (2.4 GHz), 10 (915 MHz), 1 (868 MHz)	16 (2.4 GHz), 10 (915 MHz), 1 (868 MHz)	79 40 (v4.0)	14 (2.4 GHz) 8 (5 GHz)
Network topology	Star, Mesh, and Peer-to-Peer	Star, Mesh, and Peer-to-Peer	Star, Peer-to-Peer	Star, Peer-to-Peer and Mesh	Star, Peer-to-Peer	Star, Peer-to-Peer
MAC scheme	CSMA/CA TDMA + CSMA/CA (Star Topology)	TDMA + CSMA/CA (beacon mode)	CSMA/CA (beaconless mode)	TDMA + CSMA/CA (beacon mode)	TDD	CSMA/CA + PCF
Modulation scheme	BPSK (868–915 MHz) Q-QPSK (2.4 GHz)	O-QPSK (2.4 GHz)	FSK/OOK	O-QPSK (2.4 GHz)	GFSK/DQPSK 8DPSK (optional)	BPSK, QPSK, COFDM, CCK, M-QAM
Nominal rate	250 kbps (2.4 GHz)	250 kbps (2.4 GHz)	250 kbps (2.4 GHz)	250 kbps (2.4 GHz)	1 Mbps (v1.2/v4.0)	11-65-450 (IEEE 802.11 n) Mbps
	40 kbps (915 MHz)	40 kbps (915 MHz)	40 kbps (915 MHz)	40 kbps (915 MHz)	3 Mbps (v2.0)	
	20 kbps (868 MHz)	20 kbps (868 MHz)	20 kbps (868 MHz)	20 kbps (868 MHz)	24 Mbps (v3.0)	
Power efficient mechanism	Supported	Supported	Supported	Supported	Supported	Supported
Encryption	AES128	AES128	AES128	AES128	AES64 and AES128	CCMP 128
Data authentication	MIC-32, MIC-64, MIC-128 (Shared key approach) ENC-MIC-32,	MIC-32, MIC-64, MIC-128 (Shared key approach) ENC-MIC-32,	MIC-32; MIC-64; MIC-128 (Shared key approach) ENC-MIC-32;	MIC-32; MIC-64; MIC-128 (Shared key approach) ENC-MIC-32;	Contest response scheme	4-Way handshake

(continued)

Table 2.4 (continued)

Protocols/Specifications	ZigBee	WirelessHART	MiWi Over	Isa100.11	Bluetooth	Wi-Fi IEEE 802.11 a/b/g/n/ac/i
	ENC-MIC-64, ENC-MIC-128 (encrypted key)	ENC-MIC-64, ENC-MIC-128 (encrypted key)	ENC-MIC-64; ENC-MIC-128 (encrypted key)	ENC-MIC-64; ENC-MIC-128 (encrypted key)		
Error detection	CRC16	CRC16	CRC32	CRC16	CRC32	CRC32
Life hours	100 to 1000+	Be subject to battery specifications	Be subject to battery specifications	Be subject to battery specifications	1–10	0.5–5
Distance (m)	10–300	100	20–50	100–200	10	10–100
Areas of application	Environmental monitoring	Industrial monitoring and control	Home, consumer electronics, AMR metering, automotive, industrial, and medical applications	Industrial and control market	Wireless connectivity between consumer devices such as headphones, mobile phones or laptops	Wireless LAN connectivity, broadband Internet access
Advantages	Power efficient, several application profiles (home automation, smart energy) and topology flexibility	Data Security, reliability	Flexible, cost-effective platform	Power efficient and communication security	Speed and flexibility	Speed and flexibility

802.15.4 is a standard for power efficient, low data rate and short range wireless communication between small devices. This specification only describes PHY and medium access control (MAC) layers, is currently the most widely adopted standard for WSN for short range, low data rate (250 kb/s at 2.4 GHz), and low complication. There are several protocols like ZigBee, MiWi, 6LoWPAN, WirelessHART and ISA100.11a, which are based on IEEE 802.15.4 and known as higher layer protocol. Also, these higher layer protocols describe the recommended application structure, network layer, device profile, and security facilities between other functionalities. Bluetooth is intended for applications that are mainly based on computer peripherals devices, such as wireless mouse and keyboard. Bluetooth or IEEE 802.15.1 is a standard intended to be a secure and low-priced way of data communication among supported devices, creating a PAN. ZigBee standard is one of the most preferred standard for wireless sensors network for low power, short range and moderate data rate among the sensing units involved in environment monitoring. Table 2.4 shows in detail the wireless networking protocols supported on IEEE standards, whereas Table 2.5 shows the wireless networking protocols not supported on IEEE Standards (Mendes et al. 2015). The Wireless standards presented in Table 2.5 are non-IEEE certified and are designed by research groups as well as commercial companies.

2.3.4 Local Home Gateway and Server

All the sensing data from sensing network is sent to the coordinator. This data is fed to the local computer based server. The data stored in local home gateway server is analyzed and based on the outcome of the analysis the action is performed. These data analysis is done online as well as offline. Depending on personalized requirement analysis and data extraction, the software can be written in any software language. Most of the smart home in the recent time prefers online analysis and real-time streaming of information.

By integrating sensors with devices of daily usage in the smart home, the utilization can be done in a more intelligent and smart manner. By this sensing information, home appliances and environment can be controlled from a remote location, and this is more than just turn on or off the device. Context-aware computing and ubiquitous computing are chips of the same block, and they are indispensable for each other to enhance the service quality of operation. For optimum use of server data, these two needs have to be designed and programmed correctly (Triantafyllidis et al. 2015).

Table 2.5 Wireless Network Standards not based on IEEE Standards (Mendes et al. 2015)

Protocols/Specifications	SimpliciTI	Z-Wave	Insteon	EnOcean	Wavenis	WM-Bus
ISM Bands	2.4 GHz and Sub 1 GHz	2.4 GHz, 915 MHz (USA), 868 MHz (EU)	915 MHz (USA)	315 MHz (USA) 902.875 (USA) 868 MHz (EU)	433 MHz 868 MHz (EU) 915 MHz (USA) 2.4 GHz	169 MHz 433 MHz 868 MHz
Number of radio channels	Set by the application	2	34	1	1	12
Network topology	Peer-to-peer and Star	Mesh	Dual-mesh (RF and Powerline) Peer to peer and Mesh	Peer-to-peer, Star and Mesh	Peer-to-peer and Star and Mesh	Peer-to-peer and Star
Modulation scheme	MSK	FSK, GSK, narrowband	BPSK, FSK (in ISM Band)	ASK	GFSK	FSK, GFSK, MSK, OOK, and ASK
MAC scheme	LBT (Listen before-talk)	CSMA/CA	CSMA/CA	CSMA/CA	CSMA/TDMA and CSMA/CA (otherwise)	CSMA/CA
Nominal rate	Up to 250 kbps 40 kbps	9.6 kbps (868 MHz) 40 kbps (915 MHz)	38.4 kbps	120 kbps (868.3 MHz)	From 4.8 to 100 kbps. Usually 19.2 kbps	2.4 to 100 kbps
Power saving mechanism	Supported	Supported	Supported	Supported	Supported	Supported
Encryption	Depends on the radio MAC	AES128	No	No	3DES AES128	DES AES128
Data authentication	Be subject to the radio MAC	8-bit node I.D 32-bit home I.D	24-bit pre-assigned module I.D	8/32-bit	48-bit MAC addresses	NA

(continued)

Table 2.5 (continued)

Protocols/Specifications	SimpliciTI	Z-Wave	Insteon	EnOcean	Wavenis	WM-Bus
Data integrity	Subject of radio MAC	Allocated by prime controller	CRC16	CRC8	BCH (32,21)	CRC16
Life hours (days)	Be subject to battery specifications	Be subject to battery specifications	Be subject to battery specifications	No batteries (is solar cells, electromagnetic)	Be subject to battery specifications	Be subject to battery specifications
Distance (m)	10	30	45 (outdoors)	30	200 (indoors) 1000 (outdoors)	Up to 1000
Application areas	Security such as distributed alarm and devices, energy meters and home automation& control	Residential and industrial automation and remote control functionality	Environmental sensing	Smart homes, logistics, industry, and transportation	Industrial automation and monitoring,	Smart meters (electricity, gas, water, and heat)
Advantages	Low complexity in software design	Flexible network configuration	Reliability, economical, scalability, and flexibility	Easy installation and no battery	Long lasting battery life many years	Most economical

2.4 Smart Homes Around the World

The motivations to project and develop the smart home are independent living; enhance the wellbeing, efficient use of electricity, and safety and security. The word 'smart home' is chosen for a home environment furnished with advanced technology that allows monitoring and control to its occupants, and boosts the independent living through wellness forecasting based on behavioral pattern generation and detection. To identify the difficulties and challenges towards the key performance of a smart home monitoring, we have to recognize recent and ongoing research in this field. A variety of smart home systems for ambient assisted living are proposed and developed, but there are in fact fairly few houses that apply the smart technologies. One of the main reason for this is that the complexity and varied design requirement associated with different domains of home, these domains are communications, control, entertainment, residential and living spaces. Wearable, implantable, and microsystems that can be deployed over the body area network such as Apple watch are available nowadays. The individual wears these devices or embeds in the home environment to assist someone for health care.

The leading smart homes and ADLs research projects and websites are listed as follows:

1. MIT House: http://web.mit.edu/cron/group/house_n/ (House_n 2016)
2. CASAS Smart Home: http://ailab.wsu.edu/casas/datasets.html (CASAS 2016)
3. Smart Umass: http://traces.cs.umass.edu/index.php/Smart/Smart (Smart Umass 2014)
4. Adaptive Smart House, University of Colorado: http://www.cs.colorado.edu/~mozer/index.php?dir=/Research/Projects// (Mozer 2016)
5. Duke University Smart House: http://smarthome.duke.edu/ (Duke University Smart House 2011)
6. Georgia Tech Aware Smart Home: http://awarehome.imtc.gatech.edu/ (Georgia Tech Aware Smart Home 2016)
7. Smart_Medical_Home: https://www.rochester.edu/pr/Review/V64N3/feature2.html (Marsh 2016)
8. GETALP: http://getalp.imag.fr/xwiki/bin/view/HISData/ (Fleury 2012)
9. UCI Machine Learning Repo, Smart Ambient: http://archive.ics.uci.edu/ml/datasets.html (UCI 2016)
10. Medical Automation Research Center at UVA: http://search.lib.virginia.edu/catalog/uva-lib:2165066/view#openLayer/uva-lib:2165066/1171/1771/1/1/0 (UVA 2002)

Zhang et al. proposed and implemented a role-based approach for the body area network using clean state architecture. The communication logic of Zhang's work implemented through the role of individual instead of using protocol layer model directly. Most of the roles were interdependent; the functioning of one is defined by the other. Zhang explored packet processing for the internal state by new rules and algorithm. The work of Zhang was a very initial step towards the role-based approach (Zhang et al. 2002). In another research, event-based machine to machine communication model is proposed, this model is designed for smart cities, but the applicability of this theoretical logic on the real ground is quite doubtful (Wan et al. 2012). The FamiWare project in Gámez and Fuentes (2011) was excellent effort toward the event, and role-based sensing was integrated with the internet of things to provide a context-aware solution. There are large numbers of priority, event and rule-based models designed by researchers, but none of them is integrated, customized solutions for smart home conditions (Jouppi and Rinne 2010; Sakaguchi and Kaminagayoshi 2010). Table 2.6 shows the recent ongoing research based on event and priority based smart home solutions around the world.

There are large numbers of smart home solutions available; the rest of the section discusses most relevant smart home research around the world.

- **Aware Home Research Initiative (AHRI)**: It is an academic research project based at Georgia Institute of Technology, North Avenue, Atlanta, Georgia, the USA. It is realistic, smart home laboratory, where sensor and actuators are deployed to monitor the activity of daily living. In this smart home, approach intelligence is used to facilitate the user by real-time monitoring and analysis. The permeating computing is linked with sensing information to develop an environment which is aware of occupant's needs and routine (Chen et al. 2015). Figures 2.2 and 2.3 show the deployment of heterogeneous sensing units in the AHRI.

Figure 2.3 represents the *mum's Wine Cellar*; mum's wine cellar is a smart wine rack that allows you to spot a bottle quickly in your cellar. It is based on an intuitive visual interface.

The issue with AHRi is its controlled laboratory environment, where the subject is living independently. Apart from this, Georgia Tech project included camera-based monitoring; camera-based monitoring is efficient and easy home monitoring, but most of the people in a realistic home condition still avoid this. Camera-based surveillance crosses the boundary of privacy. Figure 2.4 represents the camera installed inside the home and visual analyzed to generate activity identification of a person.

- **The Center for Advanced Studies in Adaptive (CASAS)**: CASAS Systems is a smart home project implemented at Washington State University. The CASAS smart home not only enhances the comfort of the user but also makes the cost economical. The enhancement of comfort is done by identifying, analyzing and generating a pattern from user's daily lives for intelligent and reactive automation. The cost is maintained at economic level by reducing the maintenance and power usage. The CASAS project also has the same issue that AHRI has—use of the camera (CASAS 2016).
- **GatorTech Smart House**: The GatorTech Smart House is a project run by the Mobile and Pervasive Computing Laboratory at the University of Florida. The Smart House is a persistent programmable space designed to support the elderly and disabled in their daily lives to make them more relaxed and safe. The goal of

Table 2.6 Represents limitations and drawbacks of the recent and current research on the smart home protocol development

Title Project	Category of smart home application	Wireless communication protocol and hardware	Priority or event-based	Large data handling and packet size	Proposal and implementation	Interference mitigation
Privacy-preserving data infrastructure (Fabian and Feldhaus 2014)	Appliances monitoring	RFID and Raspberry Pi	Simple, smart home application very limited so did not use any priority	Did not consider data storage issue, considered only data privacy	It is a proposal, and most of the experimental results are based on simulation	Does not support
FemiWare for ambient intelligence (Gámez and Fuentes 2011)	Ambient intelligence	TinyOS wireless nodes	Event-based software and hardware prototype	Did not consider	Proposal and real-time implementation	Does not support
Clean-state approach (Domingo 2012)	Body area network	NA	Event and priority based	Did not consider	Proposal and simulation study	Does not support
Multimodal wireless sensor network (Tunca et al. 2014)	Ambient assisted living	ZigBee protocol, XBee RF module and Arduino with different sensor	Event driven model	Did not consider	Proposal and real-time implementation	Does not support
smart home in a box of CASAS (Sprint et al. 2015)	Behavioral detection	ZigBee	Does not support	Does not consider	Proposal and realistic implementation	Does not consider
Multisensor embedded intelligent home environment monitoring system (Chen et al. 2015)	Statistical analysis	ZigBee	Does not support the event and priority-based transmission but support emergency alerts	Does not support	Proposal and realistic implementation	Does not consider

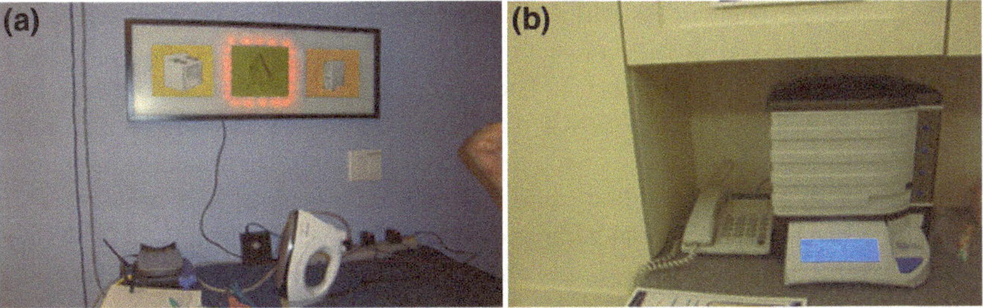

Fig. 2.2 **a** Iron Press heating system attached to the context aware arrangement, **b** medicine dosage and time reminder machine "Tabsafe" (Nazneen et al. 2015)

Fig. 2.3 Mum's wine cellar (Nazneen et al. 2015)

Fig. 2.4 Camera based monitoring (Nazneen et al. 2015)

the project is to create assisting environments that can sense themselves and their residents and create connections between the physical space, intrusion services, and remote monitoring. The smart house runs basic middleware built around the OSGi platform [OSGi] that stores service descriptions for all hardware (sensors and actuators) inside the home, in essence, turning the smart home into both as a software library and runtime environment. The middleware is designed to be easily expandable and accessible by a third party. It is divided into several application layers; the physical layer (covers all physical devices in the home), sensor platform (converts physical devices into software services), service (OS-Gi framework that maintains active services), knowledge (reasoning engine), context management (detects and registers contexts) and application (associates behavior with contexts) layers.

A SmartWave utilizes radio frequency identification (RFID) technology to program the microwave automatically for a frozen meal and plays a video helping the user sequence through the task of putting the meal safely into the microwave. If the resident does not answer or says 'no, I am not okay,' the home would call for help. Researchers at the Gator-Tech smart house are also working on cognitive prompting applications to assist an individual with mild dementia in sequencing through activities of daily living, such as washing hands or oral hygiene (Helal and Tarkoma 2015).

- **Smart home Lab at Iowa State University**: Smart Home Lab has introduced the independent living with decent daily routine through technologies of automation and security. In this project, they designed intelligence at different levels of security, such as smart doors with the camera and RFID access for access authentication. They also look at the quality of life by the intelligent fridge; this smart fridge automatically makes the list of things in stock and out of stock. To make all these they have utilized the plug-in multi-sensor board SBT30EDU, RFID Phidgets Device, X10 and Insteon are power line control technologies that enable computer monitoring and automation of appliances and Web Cameras (Heinz et al. 2012; Yang et al. 2012). The cost of the project is not economical such as the use of RFID.
- **Smart Home project**: Home monitoring at Arizona State University is the approach and methodology for pattern generation according to their object usage and movement into the urban environment. This project is almost modeling sensor occasions using a Stochastic process like Poisson processes, Continuous-time Markov Chains, etc. ASU suggested a model to generate the pattern and this model verified by CASAS smart home. Figure 2.5 represents the layout of sensor deployment and activity generation inside the home environment (Panchanathan et al. 2013).
- **AgingMO**: AgingMois smart home for elderly assisted living developed at the Tiger Place University of Missouri, USA. The primary goal of this project is elder health care and medical support by telemedicine, which is achieved by deployment of heterogeneous sensors in the home. Figure 2.6 represents the block diagram of smart home functioning from end-device to any health

emergency for an elderly assisted living while Figs. 2.7 and 2.8 present the deployment of sensing units into home ambient (Wang et al. 2014).

The AgingMo is one of the best suitable projects for elderly, but it has some drawback such as cost, limited up to elderly only and project maintenance.

- **Home Monitoring at Rochester University**: It is an alliance between researchers in the College and the School of Medicine and Dentistry—seems like an ordinary studio. However, the home-like setting is packed with technology designed to progress early recognition and anticipation of health and medical problems. The vital goal is to devise ways to help residents live longer, more healthy lives within the comforts of home. Figure 2.9 shows the sensing technology used in home monitoring project. This project is a collaboration between Advancement in Sensing Technology and Medical science, but it is under development and trial period. The project is not mature enough to implement on the real ground (Marsh 2016).
- **Mobilising Advanced Technologies for Care at Home (MATCH)**: MATCH is a combined research endeavor among the Universities of Dundee, Edinburgh, Glasgow, and Stirling. The theme and goal of this project are home care through monitoring and pattern generation of activities of occupants and voice

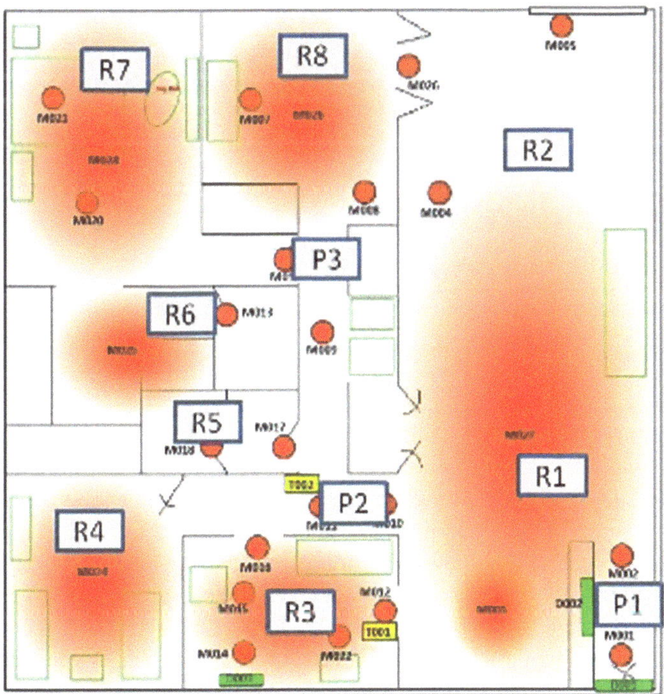

Fig. 2.5 Sensor deployment and activity recognition layout (Panchanathan et al. 2013)

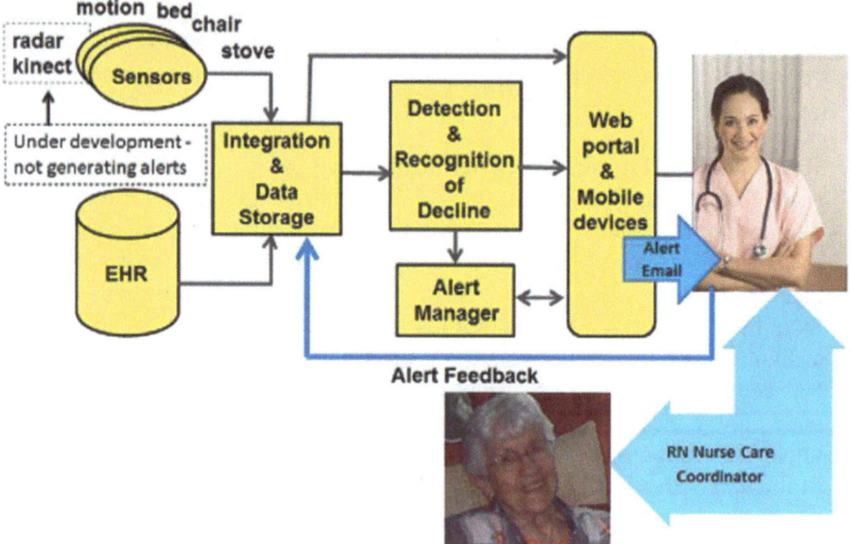

Fig. 2.6 Block diagram representation smart home for assisted living and care (Wang et al. 2014)

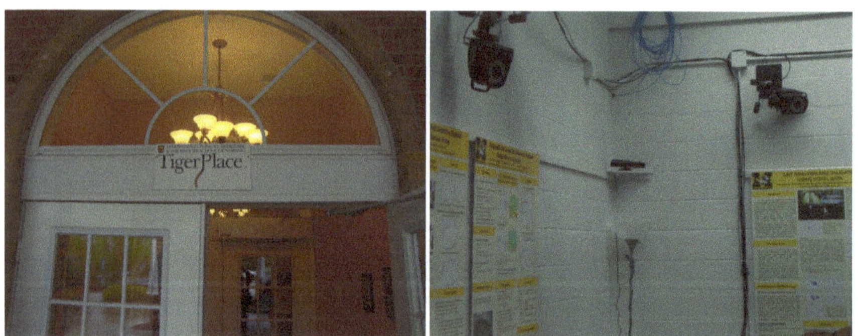

Fig. 2.7 Camera installed at Tiger place Missouri (Wang et al. 2014)

recognition. It is based on an effective reminder system, in which the tracking system assists the people by forecasting for future routine and restricts them to avoid certain activities This project started a long time ago and implemented in a laboratory environment, did not show significant progress that leads to offer the consumer. Figure 2.10 shows the home care lab bedroom; Fig. 2.11 shows kitchen and Fig. 2.12 show lounge. Whereas Fig. 2.13 shows, (a) the user is evaluating gestural input via a watch-like device and (b) front of the 'MATCH Box' developed by the University of Glasgow team (MATCH home care 2016).

- **Toyota Dream Home PAPI**: PAPI drawn up in Japan, is a commercial smart home project. The key goals of this project are to plan and realize an

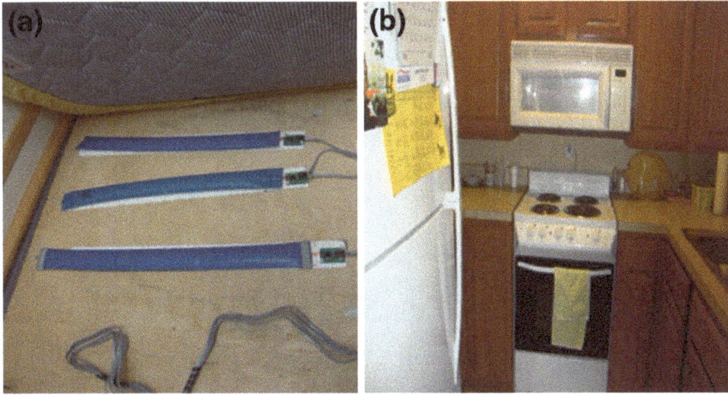

Fig. 2.8 **a** Force sensor deployed below the mattress and **b** appliances are connected to usage monitoring logic (Wang et al. 2014)

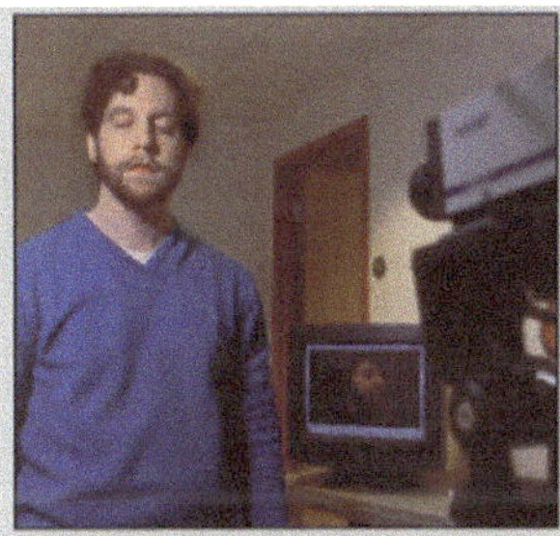

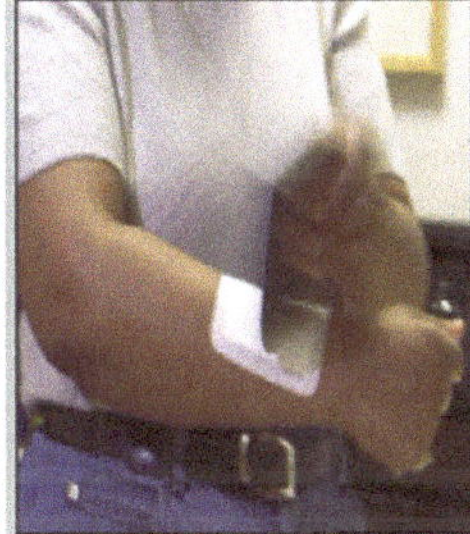

Fig. 2.9 Sensing technology applied in home monitoring (Marsh 2016)

environment-friendly, energy efficient and intelligent house design, in which the modern permeating network and computing technologies applied. In Fig. 2.14, the Ubiquitous Communicator (UC) is used as remote control all over the home. As the user moves from one room to another, the functions of the Communicator change, and it also recognizes the person and his/her favorites as

Fig. 2.10 Home care lab bedroom (MATCH home care 2016)

Fig. 2.11 Home care lab kitchen (MATCH home care 2016)

Fig. 2.12 Home care lab lounge (MATCH home care 2016)

he/she moves through the computerized living space (Sakamura 2010) (Fig. 2.15).

PAPI includes ubiquitous network, ubiquitous network/home & car, home theater, intelligent storage system, smart security system, smart auto door, a bedroom for high quality sleep, comfortable air conditioning system, blind

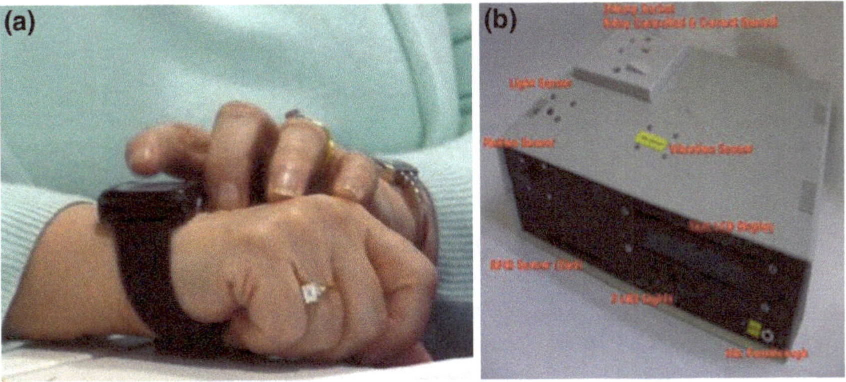

Fig. 2.13 **a** User is evaluating gestural input via a watch-like device Glasgow, © University of Glasgow and **b** front of the 'MATCH Box' developed by the University of Glasgow team (MATCH home care 2016)

Fig. 2.14 The Ubiquitous Communicator (UC) is used as remote control all over the home (Sakamura 2010)

shutter, residential elevator, eco interior furnishing, unobtrusive window screens, a power outage proof house, solar hot water supply and heating system, dye-sensitized solar cells and residential fuel cells. Figure 2.16 represents the electric car charging and emergency power supply by that same car in case of any electricity cut. Figure 2.17 shows the wide open window space for natural

Fig. 2.15 Toyota PAPI house at Japan (Sakamura 2010)

light to save electricity and big fireplace to maintain room temperature in winter (Sakamura 2010).

- **myGEKKO**: myGEKKO is a commercial smart home solution provided by an Italian company. It is a user-friendly, manufacturer-independent and multiprotocol automation and control system. myGEKKO offers us the opportunity to utilize power in a more efficient, adaptable and cost-effective manner to monitor and control the plants and devices centrally: whether at home, in enterprises or a big network. It offers home automation with fully integrated with the internet of things. For instance fire control automation system is installed to detect and handle any kind of fire hazards. Figure 2.18 shows the fireplace based room heating, and Fig. 2.19 represents the control system in case of any fire hazard due to improper use of the fireplace (myGekko 2016).

The above discussion presents a smart home research projects that are being proposed as well as implemented around the world; they do represent the diversity and diverse areas of interest that have emerged over the years. These projects can be categorized into three subcategories: University Research projects, corporate projects, and projects that have involved long-term user testing. All university projects concentrate on solving a definite research problem by using the modern technology and methods available. The Adaptive Home became an adaptive space capable of learning with the assistance of neural networks, whereas the Aware and Adaptive Home again serves as a platform for quite a few individual research projects, most

Fig. 2.16 Green car charger and power supply in the home if needed (Sakamura 2010)

of them aiming at refining our lives, connecting with family members, etc. The Smart Home presents a complex and exciting home infrastructure, but the research consists mostly of small, autonomous projects and a complete home control system with the appropriate user interface. Commercial research projects are unsurprisingly more persuaded towards promoting new products, services, and technologies. There are not that many research projects that have elaborated long-term subject tests, many laboratories have been occupied for the time being, either by researcher or volunteers, but these experiments have only lasted from a few days to months. The Duke Smart Home is also another laboratory based home environment, but in this case, their occupants are students and the laboratory their permanent home (Chen et al. 2011; Malhi et al. 2012; Chen 2011; Karia et al. 2011; Sung and Hsu 2011; Hossain and Ahmed 2012; Worringham et al. 2011; Chew et al. 2014).

This section is more focused toward the smart home approach and realization, the next section targets the most relevant data mining and machine learning methods.

Fig. 2.17 Wide open window space for natural light and a big fireplace (Sakamura 2010)

Fig. 2.18 Fireplace in the room (myGekko 2016)

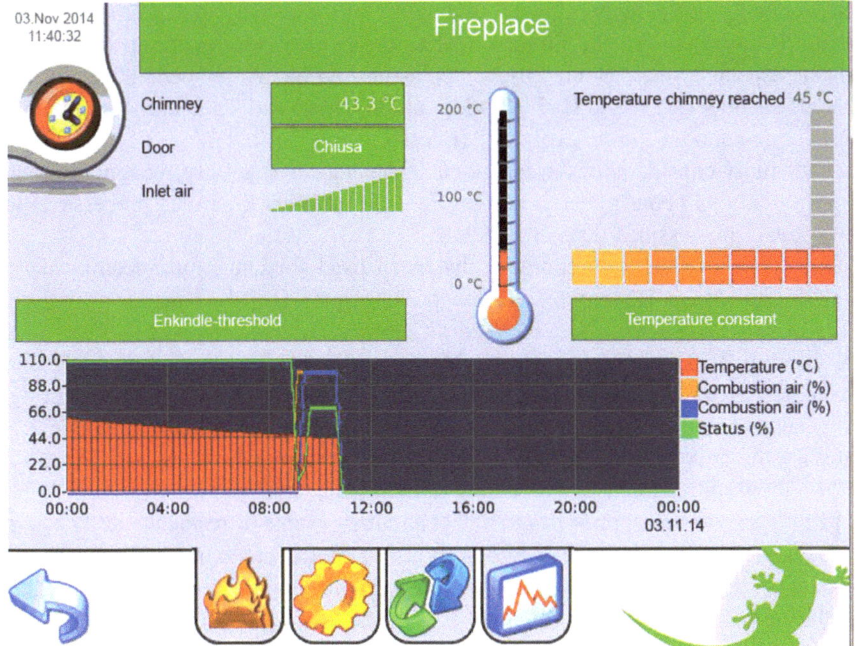

Fig. 2.19 Fire hazard control caused by the fireplace (myGekko 2016)

2.5 Activity Recognition Algorithms and Approaches

Crucial areas of interest in Assisted Living (AL) include an offer to live independent, fall prevention, emergency care and Activity of Daily Living (ADL) monitoring. The timeliness and precision of the classification of ADL events could have severe costs if inadequate, especially in the case of an emergency life risk event such as a heart attack and are consequently necessary to offer the occupant with a sense of safety and self-reliance.

ADL events are commonly classified based on the features extracted from segments of the sensing data and have a substantial role in the precision of event classification consequently. The sampling frequency (SF), segmentation method (SM) and window size (WS) play a major role in feature extraction, even though there are not many researchers who documented clear justification about parameters selection. Additionally, researchers have a tendency to overlook the essential Computational Load (CL) and Computational Resources (CR) for data classification, which is of individual interest, once data classification takes place on an embedded sensing system for real-time ADL recognition.

In general, the main deficiencies that have been stressed out by researchers most recently covering the issues of recognizing interweaved and simultaneous activities, imbalance data, multi-resident activities, realtime-activity learning, applicability and flexibility of the activity model, scalability of the activity model (Amiribesheli et al. 2015) Moreover, the limitation of using ambient sensor due to devoted to spatial aspects (Steen et al. 2013), "cold-start problem" by certain supervised models such as HMM and CRF (Cook et al. 2013a, b) and static classifier that has been none-adaptive (none-evolving) (Iglesias et al. 2012; Ordóñez et al. 2013). Added, most considerations were given to health care, activity recognition, and fewer attentions being given to activity forecast and real-time data processing (or data stream processing) (Das et al. 2010).

There are number of techniques that been used for anomaly detection, these include statistical: Histogram, GMM, probabilistic: HMM, Machine Learning: Feedforward, Recurrent Neural Network, Fuzzy System, Support Vector Data Description (SVDD), Support Vector Machine (SVM), One Class Support Vector Machine (OCSVM), Data mining: clustering, association rules, FCM clustering, K-means, agglomerative hierarchical clustering (MAHMOUD 2012). Table 2.7, presents the general classification of anomaly detection techniques, and Table 2.8 shows the traditional activity modeling approaches.

There are various approaches to detect anomaly events of residence in the smart home domain. The techniques range from how activities are extracted from sensing data, how to model normal data (profiling) when anomaly activity samples are hardly seen from past data and how to capture activity's duration. Most of the researchers claimed that occupancy's activities follow some "regular pattern" that could be learned from probabilistic models. Cardinaux et al. (2008) suggested discrete profiling strategy with Gaussian Mixture Model (GMM) to form normal data. This approach is better than previously Histogram methods (Virone et al. 2008), due to the reason that GMM could capture attribute's dependency, such as when capturing both attributes of time and duration, an activity of watching TV at 8 a.m. (time) for 5 h (duration) could be an anomaly. Behaviours are extracted by

Table 2.7 General classification of anomaly detection techniques

Algorithm	Pros	Cons	Reference
Classification (supervised learning)	• Accuracy measurement is based on confusion matrix	• As its supervised learning, it demands data instance for training • Human' biased problem in labeling data	Pulsford et al. (2011)
Clustering based (unsupervised learning)	• As it is unsupervised learning does not need data instance as a sample	• Performance is subject to the assumption about data distribution • Does not adjusted for anomaly detection	Lee et al. (2009)
Statistical based (unsupervised learning)	• Do not need data instance as a sample	• Performance depends on the assumption about data distribution • Not optimized for anomaly detection	Chehade et al. (2012)

Table 2.8 Traditional activity classification (modeling) approaches (Chen et al. 2012)

	Knowledge-driven approach (KDA)			Data-driven approaches (DDA)	
	Mining-based	Logic-based	Ontology-based	Generative	Discriminative
Model type	HMM, DBN, SVM, CRF, NN	Logical formula, e.g. plans, lattices, event trees	HMM, DBN, SVM, CRF, NN	Naïve Bayes, HMM, LDS, DBNs	NN, SVM, CRF, Decision Tree
Modelling mechanism	Information retrieval and analysis	Formal knowledge modeling	Supervised learning from datasets	Supervised learning from datasets	Supervised learning from datasets
Activity recognition method	Generative or discriminative methods	Logical inference e.g. deduction, induction	Generative or discriminative methods	Probabilistic classification	Similarity or rule-based reasoning
Advantages	No "Cold Start" problem, using multiple data sources	No, "Cold Start" problem, clear semantics on modeling & inference	Shared terms, interoperability, and reusability	Modeling uncertainty, temporal information	Modeling uncertainty, temporal information, heuristics
Disadvantages	The same problem as DDA	Weak in handling uncertainty and scalability	The same problem as DDA	"Cold start" problem, Lack of reusability & scalability	"Cold start" problem, lack of reusability & scalability

the author from the sensors through the Rule-Based algorithm, e.g., ⟨time-in, time-out, duration⟩. Some researchers claimed that Hidden Markov Model is the best suit noisy domain for the smart home. Instead of using GMM, Monekosso and Remagnino (2009) employed HMM as general profiling strategy by inputting clusters of "activity of daily livings" into the model. They have claimed that their way of unsupervised activity extraction strategy is better than traditional supervised method. Mori et al. (2007) chosen to detail the individual profiling strategy by profiling each activity's frequency of occurrence (how much the activity label occurred at a certain time) through probabilistic density model. SDLE and SDEM algorithm used by outlier detection engine SmartSifter are used here for this purpose and activities are extracted through HMM and Segmentation, in which later the alpha values are grouped into k-means clusters to form labels. Nevertheless, their approaches are judged by numerous researchers for the ineffectiveness of accurately measure anomaly behavior because same activities have different activity's duration. Example, a collection of temporal ordered sensor events: "door", "fridge", "sink", "bench", "fridge", "door" may denote the activity of cooking cereal. However, identical event collection may account for the activity of making toast if take into considering the activity's duration. Because the length of making toast is longer than preparing cereal. Kang et al. (2010) has clarified that H-HMM (Hierarchical Hidden Markov Model) is better than any old-style flat—HMM when considering the aspect of capturing activity's duration. The authors have assumed that sub-activities in the lower hierarchical level should last shorter than the coverage top main-activities (overlapping time zone) because the combination of sub-activities forms the main activity. However, Chung et al. (2008) has considered the contextual aspect of the hierarchical structure through HC-HMM (Hierarchical Context—HMM), and employed three HMM (λSC, λBR, λTR) at different context later to reason activities. The authors considered three contextual aspects of spatial, temporal and their innovative work on activity context. D-HMM is used to capture activity duration. Duong et al. (2005) have exclusively developed their own duration model with discrete Coxian distribution. Duong has claimed that their technique is better in computation time and generalization of error in comparing to any traditional method of modeling the duration, which has used "multinomial distribution" and required a large number of parameters for learning (training and classification). Hung et al. (2010) combined the advantages of HMM in handling sequential data and discriminative model of SVM to detect anomaly activities.

Forkan et al. (2015a) have pointed out two shortcomings of outdated anomaly detection system. First, the ineffectiveness of predicting future trends (anomaly ahead) that cause the failure of detecting disease's sudden attack and the second, the incorporating of single context for decision making has caused high false alarm rate. Patient's sickness could not attain the mature state, and the observation of diseases e.g. diabetes must incorporate other contexts besides the sugar level. Consequently, the authors have developed an "integrated system" by using HMM and Fuzzy Logic to detect "multiple contextual activities" and "predict" the outcome by combining all the information. In monitoring daily routine, the authors have used Gaussian Mixture Model and incorporating health status as context by correlating the status with day-to-day operations. The system could detect anomalies in activity, location, routine (collective anomaly) and changes of health status changes respectively. The first approach, the 1-class HMM is employed when anomaly data instance is unavailable, and the whole data set is used as normal data (Profile). A threshold value is defined to decide normal-anomaly boundary. During the testing phase, the log joins probability "log P(O|λ)" is used as an evaluation measure, by comparing this value with a predefined threshold value to alert anomaly. The second approach, both normal and anomaly data instances are used to model two hidden states HMM (2-HMM). During the testing phase, Viterbi algorithm (Forney 1973) is used to find the best-hidden state configuration that represents the "predicted" two hidden states, to define which activities are anomaly from the incoming, unseen testing data set (Wong et al. 2014).

Khan et al. (2012) had projected three diverse approaches of Hidden Markov Model (HMM) to detect unusual temporal events from the sensor network. The first two methods used log-likelihood threshold to detect the deviation of anomaly events from normal when the first approach modeled separately each event with

HMM and the second method modeled the whole dataset with 1-class HMM. The third approach modeled the normal events with 1-class HMM and built anomaly event model through approximation by varying the covariance of the observation anomaly events. Anomaly detection is conducted during the testing phase.

The current sub-section has documented existing data mining and machine learning methodologies applicable to ADL.

2.6 Issues of Deployment

The IEEE 802.15.4 ZigBee standard enables the wireless technologies for heterogeneous sensors at home and building environment. The performance and facilities in wireless sensing are moving from research grade to the industrial stage as a smart building monitoring and automation. In the building environment, various communication devices are operating in the ISM band that causes the degradation in performance and reliability to desired radio signal between a sensor node and data collection points. There are three phenomena interference, attenuation, and multipath. The main issue with different wireless communication technologies in the same frequency band is that most of them are not appropriately planned to be compatible with each other. There are some concerns of channel overlapping as well. The co-existence behavior of ZigBee and Wi-Fi has been documented thoroughly in different research studies (Kioumars and Tang 2011). Additionally, physical layer concerns of Wi-Fi and ZigBee coexistence have been previously considered in the IEEE 802.15.4 standard (Saleem and Rehmani 2014). Even after this, the realistic building environment performance is totally different from the ideal IEEE standard consideration. Other non-networking applications emit the electromagnetic waves such as microwave in the ISM band that disturbs the wireless communication of personal area network (PAN). ZigBee protocol is based on Direct-Sequence Spread Spectrum (DSSS) technique. These spread spectrum technologies provide some degree of immunity from interferers (Kartsakli et al. 2014).

The Physical lower layer behavior of Wi-Fi and ZigBee coexistence has been considered in different studies of WSNs design.

The existing research studies systematically consider the interference effects, but they are based on simulation approach. These studies conducted on several considerations for wireless communication environment, topology and interference sources. They used testbed assumptions such as worst case scenarios for experiments, which is totally different from realistic, practical building environment (Lazaro et al. 2013).

In their research, Zhen et al. (2007) found that the cross-technology estimation possibility of Clear Channel Assessment (CCA) between ZigBee and Wi-Fi is critical. They identified that the Wi-Fi is insensitive towards ZigBee, but ZigBee is oversensitive to Wi-Fi. The CCA range for coexistence behavior was 25 m with free space path loss model. In a similar study, Tytgat et al. (2012) presented that the

CCA range of ZigBee can cut down the collisions with Wi-Fi, but the ZigBee CCA mechanism is a bit slow to avoid all Wi-Fi channel traffic. They logged up to 85% ZigBee packet loss rate due to 802.11b contention. In the empirical data study by Sikora (2004) and extended (Sikora and Groza 2005), an initial insight into the co-existence performance of IEEE 802.15.4 ZigBee is presented. It had recorded that with the channel offset of $\sim$10 MHz the packet error rate (PER) of IEEE 802.15.4 ZigBee reduces from 92 to 30%, under Wi-Fi interference. The setup had 2 m spacing between WPAN to Wi-Fi transmitter. In a similar research, Petrova et al. (2006) briefly covered the co-existence with IEEE 802.11b/g. They performed experiments with the fixed spacing 3.5 m between Wi-Fi and WLAN and noted that for satisfactory performance of the IEEE 802.15.4 the offset frequency must be at least 7 MHz. For better co-channel rejection, the packets should be of small size. They recorded better system performance for the packets of 20 bytes as compared to the maximum packet size of 127 bytes (Iturri et al. 2012). Guo et al. (2012), in the research studies on impacts of 2.4 GHz ISM band interference on IEEE 802.15.4 WSN reliability in building environment noted that PER is ranging from 2% (no interference source) to 25% (interference source).

The other dominant interferer, microwave oven, operates at around 2.45 GHz and radiates a significant amount of electromagnetic signals. Although enclosed in a Faraday cage, it is still possible for some leakage to occur around the doors and opening. This gets worse over time as mechanical abuse or simple 'wear and tear' causes door seals to become less effective. For these reasons, microwave ovens are a potential source of interference for WPANs. In the study by Lee et al. (2014) the received PER is 12% for the transmitter node to coordinator spacing 6 m and microwave to receiver distance 0.5 m. This influence is becoming more decisive because the number of present applications based mainly on ZigBee wireless sensor networks functioning in the 2.4 GHz band in indoor scenarios is promptly growing (e.g., home automation and monitoring, energy management, health monitoring and lighting). Recently, mobile technology company Qualcomm has planned to build the methodology to utilize unregulated and unlicensed 2.4 GHz band for mobile (Qualcomm 2016). Apart from these interferes, Bluetooth and CAM radios are other devices that operate in ISM band and also affect reasonably (Cuomo et al. 2013).

In realistic building condition, the radio signal usually encounters some household objects in its endeavor of transmission and experiences further attenuation subject to the absorption characteristics of objects. The propagating electromagnetic signal in the building environment undergoes three primary physical modes that introduce attenuation: reflection, diffraction, and scattering. There are many different objects including mobile, stationary and transient objects in the indoor environment that cause the loss of RF energy. Higher frequencies attenuate much quicker than lower frequencies, similar to free space propagation loss. Path loss, as well as attenuation loss of radio signals, occurs with distance. Also, the amount of attenuation varies with the frequency of RF signal and the obstacle material type and density (Fan and Letaief 2009).

Furthermore, attenuation is directly linked with another parameter that is multi-path or direction of arrival (DOA). When a transmitter transmits its radio signal simultaneously in all directions, the signal usually passes through many paths to reach the receiver. In each and every path, the radio signal interacts communication environment ambiguously and reaches the receiver with some delay, and may be a change in phase and frequency takes place. If the signals received at the receiver are in phase, then they produce the constructive interference. In case the signals are out of phase signals, they cause a loss in signal strength as they produce destructive interference. The spread of this delay is known as delay spread, and the attenuation caused by it is called multipath fading. The multipath fading is mainly divided into two parts. First, when the obstacles are large, and static is known as large-scale fading, slow-fading, or shadowing. Second, when the obstacles are small and transient is referred to as small scale, fast fading, and scattering. The variation in frequency is known as Doppler spreading (Akyildiz et al. 2002). Zhang et al. (2011), proposed mathematical models with the primary parameters spacing, RSSI and antenna orientation for the indoor environment. In their study, they recorded packet delivery ratio (PDR) 95% for the spacing that was less than 10 m through the thick wall at 0 dBm and for the same setup the link quality indication (LQI) was less than 80%.

2.7 Large Data Handling

Ambient assisted living (AAL) system consists of heterogeneous sensors and devices which generate huge amounts of patient-specific unstructured raw data every day. Due to the variety of sensing units and other system equipment, the captured data also have extensive variations. These data elements can be from a few bytes of a numerical value to several gigabytes of the video stream (Forkan et al. 2015b; Dijck 2014). For instance, if we assume a single AAL home monitoring system produces 50 kB data every second on average so that it will become 1.46 TB in one year (Wigan and Clarke 2013; Demchenko et al. 2013). Including these dynamically produced continuous monitoring data, there are also enormous amounts of persistent data such as occupant's profile, medical records, disease histories and social contacts. If we want to store all these data and occupant's histories to forecast any anomaly correctly, the size of data will be zettabytes in upcoming years. Such concerns necessitate the development of event and priority based infrastructure which provide intelligent sampling for transmission control (Michael and Miller 2013; Tallon 2013; Bergelt et al. 2014).

According to IBM data researchers, big data can be described in four dimensions: volume, variety, velocity, and veracity ("the 4 V's") (Malik 2013). The Integrated framework based on Wellness Protocol also fulfills these four V's because of event and priority based data transmission. The predictive analyses over large historical data require significant computational time and resources, so the data should be handled properly to reduce this cost (Tsai et al. 2014). This behavioral pattern generation and forecasting analysis provide robust solutions for disease prevention and also suggests healthy lifestyle for an individual. This also simplifies the tasks of caregivers, healthcare professionals, and doctors by evaluating the roots of any anomalous situation at an early stage and improving the quality of life of an Individual (Chang et al. 2014; Fang et al. 2014).

2.8 Introducing Internet of Things, Web of Things and Cloud Computing

A few years ago, the advancement of the Internet of Things (IoT) was considered as a next generation technology. By the advances in digital technology, IoT is no longer just a buzzword—rather it is realistic and certain technology. The prediction for future development and investment is promising: Cisco predicts that the end of 2015 will link 25 billion devices, and 50 billion by 2020. In the 21st century, more devices than persons are associated with the internet—over 12.5 billion devices in 2010. On the IoT prototype, some objects available in the environment will be on the communications network and can be reachable in any part of the world. Radio communication and sensor network expertise will take shape to fulfill the new challenges, among which are the information and communication technology systems to be deployed in the ambient. Unquestionably, the key strength of the IoT approach is the surprising impression it will make on numerous aspects of the daily activities of consumers. IoT is documented by the US National Intelligence Council (NIC) in the table of six "Disruptive Civil Technologies" with potential influences on US national power. NIC forecasts that by the year 2025, internet-based nodes may be present in the whole thing—from food packages to healthcare medicines and measuring instruments, from miniature household objects to furniture and more (Cook 2012; Atzori et al. 2010).

In the recent years, the sensor data have crossed the borders of the local home gateway server and reached to remote access over the internet. This sensing information is analyzed and treated at a higher abstraction level for decision making through data mining and machine learning procedures and models. The recently developed field that integrates the sensing data with internet facilities for remote access is the semantic sensor web (Aijaz 2015).

At the micro level, the Smart City approach includes the Smart Home scenario; but the aspects of data analysis, data mining, and machine learning define the particular facilities of Smart Home based AAL solutions (Yang et al. 2014).

There is still a series of queries that arise as to how the IoT applications would develop and be installed in smart homes and buildings, so, there are several new problems regarding networking characteristics. These questions are linked to security, reliability, complexity, discoverability and interoperability. Widespread acceptance of such recent technology could lead to future risk. Undeniably, it is clear that usage of daily objects and other activities, which are linked to IoT, could trigger distribution of information and generate security concerns. There is a large number of companies who have financed big money in research and development on IoT, such as smart sensors (Bosch, STMicroelectronics, IoT ignition lab Moscow), embedded systems (ARM, Infineon), network vendors (Ericsson), software (Atos, SAP, Microsoft Azure), telecoms (Orange) and application integrators (Siemens, Philips) (Chen et al. 2014).

Extending the concept of the IoT, the Web of Things is an idea where daily usage devices and sensors are connected by fully integrating them into the Web. The WoT introduces many benefits in the ubiquitous computing society.

2.9 Conclusion

Various recent and existing research based on smart home monitoring conducted by different researchers across the world have been documented in this chapter. Though these existing research findings do not answer necessary questions related to security, reliability, complexity, discoverability, cost and interoperability. Most of the smart home systems either are a proposal or implemented systematically into a controlled environment. Most of the systems are based on probabilistic methodologies; ultimately it raises a severe problem of false alarm. Additionally, these existing approaches are based on offline analysis, which is not acceptable in the case of AAL where the real-time analysis of streaming data is inevitable.

Moreover, none of the AAL researchers gave emphasis on the issue of interference and large data, which is significant for the system performance, handling, and maintenance.

References

B. Depaula, (2014, June 23). *Where in the World Do People Live Alone?*. Retrieved Nov 18, 2016, from http://blogs.psychcentral.com/single-at%20heart/2014/06/where-in-the-world-do-people-live-alone/.

E. Badger, (2013, August 27). *A Brief History of How Living Alone Came to Seem Totally Normal.* Retrieved Nov 10, 2016, from http://www.citylab.com/housing/2013/08/brief-history-how-living-alone-came-seem-totally-normal/6689/.

M. Marmot and R. Bell, "Fair society, healthy lives," *Public Health,* vol. 126, pp. S4–S10, 2012.

K. Kinsella, J. Beard, and R. Suzman, "Can Populations Age Better, Not Just Live Longer?," *Generations,* vol. 37, pp. 19–26, 2013.

K. Raworth, "A safe and just space for humanity: can we live within the doughnut," *Oxfam Policy and Practice: Climate Change and Resilience,* vol. 8, pp. 1–26, 2012.

P. Novitzky, A. F. Smeaton, C. Chen, K. Irving, T. Jacquemard, F. O'Brolcháin, *et al.*, "A review of contemporary work on the ethics of ambient assisted living technologies for people with dementia," *Science and engineering ethics,* vol. 21, pp. 707–765, 2015.

I. Bisio, F. Lavagetto, M. Marchese, and A. Sciarrone, "Smartphone-centric ambient assisted living platform for patients suffering from co-morbidities monitoring," *Communications Magazine, IEEE,* vol. 53, pp. 34–41, 2015.

K. Wang, Y. Shao, L. Shu, G. Han, and C. Zhu, "LDPA: a local data processing architecture in ambient assisted living communications," *Communications Magazine, IEEE,* vol. 53, pp. 56–63, 2015.

P. M. Butala, Y. Zhang, T. D. Little, and R. C. Wagenaar, "Wireless system for monitoring and real-time classification of functional activity," in *Communication Systems and Networks (COMSNETS), 2012 Fourth International Conference on*, Bangalore India, 2012, pp. 1–5.

H. B. Stauffer, "Smart enabling system for home automation. Consumer Electronics," *IEEE Transactions on,* vol. 37, pp. xxix–xxxv, 1991.

P. Wacks, "The impact of home automation on power electronics," in *Applied Power Electronics Conference and Exposition, 1993. APEC'93. Conference Proceedings 1993., Eighth Annual,* San Diego, CA, 07–11 Mar 1993, pp. 3–9.

V. Ricquebourg, D. Menga, D. Durand, B. Marhic, L. Delahoche, and C. Loge, "The smart home concept: our immediate future," in *E-Learning in Industrial Electronics, 2006 1ST IEEE International Conference on*, Hammamet, 18–20 Dec, 2006, pp. 23–28.

H. Ghayvat, S. Mukhopadhyay, and X. Gui, "Issues and mitigation of interference, attenuation and direction of arrival in IEEE 802.15. 4/ZigBee to wireless sensors and networks based smart building," *Measurement,* vol. 86, pp. 209–226, 2016.

A. Leone, G. Rescio, and P. Siciliano, "A Near Field Communication-Based Platform for Mobile Ambient Assisted Living Applications," in *Ambient Assisted Living*, ed: Springer, 2015, pp. 125–132.

D. He and S. Zeadally, "Authentication protocol for an ambient assisted living system," *Communications Magazine, IEEE,* vol. 53, pp. 71–77, 2015.

M. S. Dias, E. Vilar, F. Sousa, A. Vasconcelos, F. M. Pinto, N. Saldanha, *et al.*, "A Living Labs Approach for Usability Testing of Ambient Assisted Living Technologies," in *Design, User Experience, and Usability: Design Discourse*, ed: Springer, 2015, pp. 167–178.

D. W. Seo, H. Kim, J. S. Kim, and J. Y. Lee, "Hybrid reality-based user experience and evaluation of a context-aware smart home," *Computers in Industry*, vol. 76, pp. 11–23, 2016.

A. Rodrigues, T. Camilo, J. Sa Silva, F. Boavida, M. Silva, N. Blanco, et al., "Hermes: A versatile platform for wireless embedded systems," *in World of Wireless, Mobile and Multimedia Networks (WoWMoM)*, 2012 IEEE International Symposium on a, San Francisco, CA, 25–28 June, 2012, pp. 1-9.

S. Dengler, A. Awad, and F. Dressler, "Sensor/actuator networks in smart homes for supporting elderly and handicapped people," *in Advanced Information Networking and Applications Workshops, 2007*, AINAW'07. 21st International Conference on, Niagara Falls, Ont. 21–23 May, 2007, pp. 863–868.

M. Wang, G. Zhang, C. Zhang, J. Zhang, and C. Li, "An IoT-based appliance control system for smart homes," in *Intelligent Control and Information Processing (ICICIP), 2013 Fourth International Conference on*, Beijing, 9–11 June 2013, pp. 744–747.

J. Byun, B. Jeon, J. Noh, Y. Kim, and S. Park, "An intelligent self-adjusting sensor for smart home services based on ZigBee communications," *Consumer Electronics, IEEE Transactions on,* vol. 58, pp. 794–802, 2012.

T. D. Mendes, R. Godina, E. M. Rodrigues, J. C. Matias, and J. P. Catalão, "Smart home communication technologies and applications: Wireless protocol assessment for home area network resources," *Energies,* vol. 8, pp. 7279–7311, 2015.

A. Triantafyllidis, C. Velardo, T. Chantler, S. A. Shah, C. Paton, R. Khorshidi, *et al.*, "Effect of daily remote monitoring on pacemaker longevity: a retrospective analysis," *Heart Rhythm,* vol. 12, pp. 330–337, 2015.

House_n, (2016, Nov 12). *House_n Materials and Media.* Retrieved Nov 18, 2016, from http://web.mit.edu/cron/group/house_n/publications.html.

CASAS, (2016, Nov 10). *CASAS.* Retrieved Nov 19, 2016, from http://ailab.wsu.edu/casas.

Smart Umass, (2014, Feb 21). *Trace Umass*. Retrieved Nov 19, 2016, from http://traces.cs.umass.edu/indeX.php/Smart/Smart.

M.C. Mozer, (2016, Nov 10). *Adaptive Smart House University of Colorado*. Retrieved Nov 10, 2016, from http://www.cs.colorado.edu/~mozer/index.php?dir=/Research/Projects//.

Duke University Smart House, (2011). *Where ideas for smart living started*. Retrieved Nov 10, 2016, from http://smarthome.duke.edu/.

Georgia Tech Aware Smart Home, (2016). *Aware research home initiative*. Retrieved Nov 10, 2016, from http://awarehome.imtc.gatech.edu/.

J. Marsh, (2016). *Smart_Medical_Home*. Retrieved Nov 10, 2016, from https://www.rochester.edu/pr/Review/V64N3/feature2.html.

J. Fleury, (2012, Jan 16). *GETALP*. Retrieved Nov 10, 2016, from http://getalp.imag.fr/xwiki/bin/view/HISData/.

UCI, (2016). *Machine Learning Repo, Smart Ambient*. Retrieved Nov 10, 2016, from http://archive.ics.uci.edu/ml/datasets.html.

UVA, (2002, June). *Medical Automation Research Center at UVA*. Retrieved Nov 10, 2016, from http://search.lib.virginia.edu/catalog/uva-lib:2165066/view#openLayer/uva-lib:2165066/1171/1771/1/1/0.

L. Zhang, G.-J. Ahn, and B.-T. Chu, "A role-based delegation framework for healthcare information systems," in *Proceedings of the seventh ACM symposium on Access control models and technologies*, Monterey, California, USA, June 3–4, 2002, pp. 125–134.

J. Wan, D. Li, C. Zou, and K. Zhou, "M2M communications for smart city: an event-based architecture," in *Computer and Information Technology (CIT), 2012 IEEE 12th International Conference on*, Chengdu, 27–29 Oct, 2012, pp. 895–900.

N. Gámez and L. Fuentes, "FamiWare: a family of event-based middleware for ambient intelligence," *Personal and Ubiquitous Computing*, vol. 15, pp. 329–339, 2011.

J. Jouppi and J. Rinne, "Transfer of packet data to wireless terminal," Google Patents, US7643456 B, Jan 5, 2010.

H. Sakaguchi and T. Kaminagayoshi, "Role play system," Google Patents, US7794315 B2, Sep 14, 2010.

B. Fabian and T. Feldhaus, "Privacy-preserving data infrastructure for smart home appliances based on the Octopus DHT," *Computers in Industry*, vol. 65, pp. 1147–1160, 2014.

M. C. Domingo, "Throughput efficiency in body sensor networks: A clean-slate approach," *Expert Systems with Applications*, vol. 39, pp. 9743–9754, 2012.

C. Tunca, H. Alemdar, H. Ertan, O. D. Incel, and C. Ersoy, "Multimodal wireless sensor network-based ambient assisted living in real homes with multiple residents," *Sensors*, vol. 14, pp. 9692–9719, 2014.

G. Sprint, D. J. Cook, and D. L. Weeks, "Toward Automating Clinical Assessments: A Survey of the Timed Up and Go," *Biomedical Engineering, IEEE Reviews in*, vol. 8, pp. 64–77, 2015.

S.-L. Chen, S.-K. Chang, and Y.-Y. Chen, "Development of a multisensor embedded intelligent home environment monitoring system based on digital signal processor and Wi-Fi," *International Journal of Distributed Sensor Networks*, vol. 2015, p. 88, 2015.

N. Nazneen, A. Rozga, C. J. Smith, R. Oberleitner, G. D. Abowd, and R. I. Arriaga, "A Novel System for Supporting Autism Diagnosis Using Home Videos: Iterative Development and Evaluation of System Design," *JMIR mHealth and uHealth*, vol. 3, 2015.

S. Helal and S. Tarkoma, "Smart Spaces [Guest editors' introduction]," *Pervasive Computing, IEEE*, vol. 14, pp. 22–23, 2015.

M. Heinz, P. Martin, J. A. Margrett, M. Yearns, W. Franke, H. I. Yang, *et al.*, "Perceptions of technology among older adults," *Journal of Gerontological Nursing*, 2012.

H.-I. Yang, R. Babbitt, J. Wong, and C. K. Chang, "A framework for service morphing and heterogeneous service discovery in smart environments," in *Impact Analysis of Solutions for Chronic Disease Prevention and Management*, ed: Springer, 2012, pp. 9–17.

S. Panchanathan, T. McDaniel, and V. N. Balasubramanian, "An interdisciplinary approach to the design, development and deployment of person-centered accessible technologies," in *Recent*

Trends in Information Technology (ICRTIT), 2013 International Conference on, Chennai, India, 25–27 July 2013, pp. 750–757.

F. Wang, M. Skubic, M. Rantz, and P. E. Cuddihy, "Quantitative gait measurement with pulse-doppler radar for passive in-home gait assessment," *Biomedical Engineering, IEEE Transactions on,* vol. 61, pp. 2434–2443, 2014.

MATCH home care, (2016). *Mobilizing advanced technologies for care at home.* Retrieved Nov 10, 2016, from http://www.match-project.org.uk/resources/project_album.html.

K. Sakamura, (2010). *Toyota Papi Dream House.* Retrieved Nov 10, 2016, from http://tronweb.super-nova.co.jp/toyotadreamhousepapi.html.

myGekko, (2016). *myGekko personalized, user friendly and effective.* Retrieved Nov 10, 2016, from https://www.my-gekko.com/en/.

B.-R. Chen, S. Patel, T. Buckley, R. Rednic, D. J. McClure, L. Shih, *et al.*, "A web-based system for home monitoring of patients with Parkinson's disease using wearable sensors," *Biomedical Engineering, IEEE Transactions on,* vol. 58, pp. 831–836, 2011.

K. Malhi, S. C. Mukhopadhyay, J. Schnepper, M. Haefke, and H. Ewald, "A Zigbee-based wearable physiological parameters monitoring system," *Sensors Journal, IEEE,* vol. 12, pp. 423–430, 2012.

C.-M. Chen, "Web-based remote human pulse monitoring system with intelligent data analysis for home health care," *Expert Systems with Applications,* vol. 38, 2011.

D. C. Karia, V. Adajania, M. Agrawal, and S. Dandekar, "Embedded web server application based automation and monitoring system," in *Signal Processing, Communication, Computing and Networking Technologies (ICSCCN), 2011 International Conference on*, Thuckafay, 21–22 July, 2011, pp. 634-637.

W.-T. Sung and Y.-C. Hsu, "Designing an industrial real-time measurement and monitoring system based on embedded system and ZigBee," *Expert Systems with Applications,* vol. 38, pp. 4522–4529, 2011.

M. A. Hossain and D. T. Ahmed, "Virtual caregiver: an ambient-aware elderly monitoring system," *Information Technology in Biomedicine, IEEE Transactions on,* vol. 16, pp. 1024–1031, 2012.

C. Worringham, A. Rojek, and I. Stewart, "Development and feasibility of a smartphone, ECG and GPS based system for remotely monitoring exercise in cardiac rehabilitation," *PloS one,* vol. 6, p. e14669, 2011.

E. Y. Chew, T. E. Clemons, S. B. Bressler, M. J. Elman, R. P. Danis, A. Domalpally, *et al.*, "Randomized trial of a home monitoring system for early detection of choroidal neovascularization home monitoring of the Eye (HOME) study," *Ophthalmology,* vol. 121, pp. 535–544, 2014.

F. Palumbo, J. Ullberg, A. Štimec, F. Furfari, L. Karlsson, and S. Coradeschi, "Sensor network infrastructure for a home care monitoring system," *Sensors,* vol. 14, pp. 3833–3860, 2014.

W. Enck, P. Gilbert, S. Han, V. Tendulkar, B.-G. Chun, L. P. Cox, *et al.*, "TaintDroid: an information-flow tracking system for realtime privacy monitoring on smartphones," *ACM Transactions on Computer Systems (TOCS),* vol. 32, p. 5, 2014.

J. Han, C.-S. Choi, W.-K. Park, I. Lee, and S.-H. Kim, "Smart home energy management system including renewable energy based on ZigBee and PLC," *Consumer Electronics, IEEE Transactions on,* vol. 60, pp. 198–202, 2014.

N. K. Suryadevara, S. C. Mukhopadhyay, S. D. T. Kelly, and S. P. S. Gill, "WSN-based smart sensors and actuator for power management in intelligent buildings," *Mechatronics, IEEE/ASME Transactions on,* vol. 20, pp. 564–571, 2015.

S. C. Mukhopadhyay, "Wearable sensors for human activity monitoring: A review," *Sensors Journal, IEEE,* vol. 15, pp. 1321–1330, 2015.

M. Amiribesheli, A. Benmansour, and A. Bouchachia, "A review of smart homes in healthcare," *Journal of Ambient Intelligence and Humanized Computing,* pp. 1–23, 2015.

E.-E. Steen, T. Frenken, M. Eichelberg, M. Frenken, and A. Hein, "Modeling individual healthy behavior using home automation sensor data: Results from a field trial," *Journal of Ambient Intelligence and Smart Environments,* vol. 5, pp. 503–523, 2013.

D. J. Cook, N. C. Krishnan, and P. Rashidi, "Activity discovery and activity recognition: A new partnership," *Cybernetics, IEEE Transactions on,* vol. 43, pp. 820–828, 2013a.

D. J. Cook, A. S. Crandall, B. L. Thomas, and N. C. Krishnan, "CASAS: A smart home in a box," *Computer,* vol. 46, 2013b.

J. A. Iglesias, P. Angelov, A. Ledezma, and A. Sanchis, "Creating evolving user behavior profiles automatically," *Knowledge and Data Engineering, IEEE Transactions on,* vol. 24, pp. 854–867, 2012.

F. J. Ordóñez, J. A. Iglesias, P. De Toledo, A. Ledezma, and A. Sanchis, "Online activity recognition using evolving classifiers," *Expert Systems with Applications,* vol. 40, pp. 1248–1255, 2013.

B. Das, C. Chen, N. Dasgupta, D. J. Cook, and A. M. Seelye, "Automated prompting in a smart home environment," in *Data Mining Workshops (ICDMW), 2010 IEEE International Conference on*, 2010, pp. 1045–1052.

S. M. MAHMOUD, "Thesis - Identification and Prediction of Abnormal Behaviour Activities of Daily Living in Intelligent Environments" 2012.

R. M. Pulsford, M. Cortina-Borja, C. Rich, F.-E. Kinnafick, C. Dezateux, and L. J. Griffiths, "Actigraph accelerometer-defined boundaries for sedentary behaviour and physical activity intensities in 7 year old children," *PloS one,* vol. 6, p. e21822, 2011.

M.-S. Lee, J.-G. Lim, K.-R. Park, and D.-S. Kwon, "Unsupervised clustering for abnormality detection based on the tri-axial accelerometer," in *ICCAS-SICE, 2009*, 2009, pp. 134–137.

N. H. Chehade, P. Ozisik, J. Gomez, F. Ramos, and G. Pottie, "Detecting stumbles with a single accelerometer," in *Engineering in Medicine and Biology Society (EMBC), 2012 Annual International Conference of the IEEE*, 2012, pp. 6681–6686.

L. Chen, J. Hoey, C. D. Nugent, D. J. Cook, and Z. Yu, "Sensor-based activity recognition," *Systems, man, and cybernetics, Part C: Applications and reviews, IEEE Transactions on,* vol. 42, pp. 790–808, 2012.

F. Cardinaux, S. Brownsell, M. Hawley, and D. Bradley, "Modelling of behavioural patterns for abnormality detection in the context of lifestyle reassurance," in *Progress in Pattern Recognition, Image Analysis and Applications*, ed: Springer, 2008, pp. 243–251.

G. Virone, M. Alwan, S. Dalal, S. W. Kell, B. Turner, J. A. Stankovic, *et al.*, "Behavioral patterns of older adults in assisted living," *Information Technology in Biomedicine, IEEE Transactions on,* vol. 12, pp. 387–398, 2008.

D. N. Monekosso and P. Remagnino, "Anomalous behavior detection: Supporting independent living," *Intelligent Environments,* pp. 33–48, 2009.

T. Mori, A. Fujii, M. Shimosaka, H. Noguchi, and T. Sato, "Typical behavior patterns extraction and anomaly detection algorithm based on accumulated home sensor data," in *Future Generation Communication and Networking (FGCN 2007)*, 2007, pp. 12–18.

W. Kang, D. Shin, and D. Shin, "Detecting and predicting of abnormal behavior using hierarchical markov model in smart home network," in *Industrial Engineering and Engineering Management (IE&EM), 2010 IEEE 17Th International Conference on*, 2010, pp. 410–414.

P.-C. Chung and C.-D. Liu, "A daily behavior enabled hidden Markov model for human behavior understanding," *Pattern Recognition,* vol. 41, pp. 1572–1580, 2008.

T. V. Duong, H. H. Bui, D. Q. Phung, and S. Venkatesh, "Activity recognition and abnormality detection with the switching hidden semi-markov model," in *Computer Vision and Pattern Recognition, 2005. CVPR 2005. IEEE Computer Society Conference on*, 2005, pp. 838–845.

Y.-X. Hung, C.-Y. Chiang, S. J. Hsu, and C.-T. Chan, "Abnormality detection for improving elder's daily life independent," in *Aging Friendly Technology for Health and Independence*, ed: Springer, 2010, pp. 186–194.

A. R. M. Forkan, I. Khalil, Z. Tari, S. Foufou, and A. Bouras, "A context aware approach for long-term behavioural change detection and abnormality prediction in ambient assisted living," *Pattern Recognition,* vol. 48, pp. 628–641, 2015a.

G. D. Forney Jr, "The viterbi algorithm," *Proceedings of the IEEE,* vol. 61, pp. 268–278, 1973.

K. B.-Y. Wong, T. Zhang, and H. Aghajan, "Data Fusion with a Dense Sensor Network for Anomaly Detection in Smart Homes," in *Human Behavior Understanding in Networked Sensing*, ed: Springer, 2014, pp. 211–237.

S. S. Khan, M. E. Karg, J. Hoey, and D. Kulic, "Towards the detection of unusual temporal events during activities using hmms," in *Proceedings of the 2012 ACM Conference on Ubiquitous Computing*, 2012, pp. 1075–1084.

A. H. Kioumars and L. Tang, "ATmega and XBee-based wireless sensing," in *Automation, Robotics and Applications (ICARA), 2011 5th International Conference on*, Wellington, 6–8 Dec, 2011, pp. 351–356.

Y. Saleem and M. H. Rehmani, "Primary radio user activity models for cognitive radio networks: A survey," *Journal of Network and Computer Applications,* vol. 43, pp. 1–16, 2014.

E. Kartsakli, A. S. Lalos, A. Antonopoulos, S. Tennina, M. D. Renzo, L. Alonso, *et al.*, "A survey on M2M systems for mHealth: a wireless communications perspective," *Sensors,* vol. 14, pp. 18009–18052, 2014.

A. Lazaro, D. Girbau, P. Moravek, and R. Villarino, "A study on localization in wireless sensor networks using frequency diversity for mitigating multipath effects," *Elektronika ir Elektrotechnika,* vol. 19, pp. 82–87, 2013.

B. Zhen, H.-B. Li, S. Hara, and R. Kohno, "Clear channel assessment in integrated medical environments," *EURASIP Journal on Wireless Communications and Networking,* vol. 2008, pp. 1–8, 2007.

L. Tytgat, O. Yaron, S. Pollin, I. Moerman, and P. Demeester, "Avoiding collisions between IEEE 802.11 and IEEE 802.15. 4 through coexistence aware clear channel assessment," *EURASIP Journal on Wireless Communications and Networking,* vol. 2012, pp. 1–15, 2012.

A. Sikora, "Compatibility of IEEE802. 15.4 (Zigbee) with IEEE802. 11 (WLAN), Bluetooth, and Microwave Ovens in 2.4 GHz ISM-Band," Test Report. Steinbeis-Transfer Center, University of Cooperative Education, Lörrach, 2004.

A. Sikora and V. F. Groza, "Coexistence of IEEE802. 15.4 with other Systems in the 2.4 GHz-ISM-Band," in Instrumentation and Measurement Technology Conference, 2005. IMTC 2005. Proceedings of the IEEE, Ottawa, Ont. 16–19 May, 2005, pp. 1786–1791.

M. Petrova, J. Riihijärvi, P. Mähönen, and S. Labella, "Performance study of IEEE 802.15. 4 using measurements and simulations," in *Wireless communications and networking conference, 2006. WCNC 2006. IEEE*, Las Vegas, NV, 3–6 April, 2006, pp. 487–492.

P. L. Iturri, J. A. Nazábal, L. Azpilicueta, P. Rodriguez, M. Beruete, C. Fernández-Valdivielso, *et al.*, "Impact of high power interference sources in planning and deployment of wireless sensor networks and devices in the 2.4 GHz frequency band in heterogeneous environments," *Sensors,* vol. 12, pp. 15689–15708, 2012.

W. Guo, W. M. Healy, and M. Zhou, "Impacts of 2.4-GHz ISM band interference on IEEE 802.15. 4 wireless sensor network reliability in buildings," *Instrumentation and Measurement, IEEE Transactions on,* vol. 61, pp. 2533–2544, 2012.

S. Lee, H. Shin, R. Ha, and H. Cha, "IEEE 802.15. 4a CSS-based mobile object locating system using sequential Monte Carlo method," *Computer Communications,* vol. 38, pp. 13–25, 2014.

Qualcomm, (2016). *Stay informed Qualcomm.* Retrieved Nov 10, 2016, from https://www.qualcomm.com/invention/technologies/1000x/spectrum.

F. Cuomo, A. Abbagnale, and E. Cipollone, "Cross-layer network formation for energy-efficient IEEE 802.15. 4/ZigBee Wireless Sensor Networks," *Ad Hoc Networks,* vol. 11, pp. 672–686, 2013.

G. Li, P. Fan, and K. B. Letaief, "Rayleigh fading networks: a cross-layer way," *Communications, IEEE Transactions on,* vol. 57, pp. 520–529, 2009.

I. F. Akyildiz, W. Su, Y. Sankarasubramaniam, and E. Cayirci, "A survey on sensor networks," *Communications magazine, IEEE,* vol. 40, pp. 102–114, 2002.

R.-B. Zhang, J.-G. Guo, F.-H. Chu, and Y.-C. Zhang, "Environmental-adaptive indoor radio path loss model for wireless sensor networks localization," *AEU-International Journal of Electronics and Communications,* vol. 65, pp. 1023–1031, 2011.

A. Forkan, I. Khalil, A. Ibaida, and Z. Tari, “BDCaM: Big Data for Context-aware Monitoring-A Personalized Knowledge Discovery Framework for Assisted Healthcare,” 2015b.

J. Van Dijck, “Datafication, dataism and dataveillance: Big Data between scientific paradigm and ideology,” *Surveillance & Society,* vol. 12, p. 197, 2014.

M. R. Wigan and R. Clarke, “Big data's big unintended consequences,” *Computer,* vol. 46, pp. 46–53, 2013.

Y. Demchenko, P. Grosso, C. De Laat, and P. Membrey, “Addressing big data issues in scientific data infrastructure,” in *Collaboration Technologies and Systems (CTS), 2013 International Conference on,* San Diego, CA, 20–24 May, 2013, pp. 48–55.

K. Michael and K. Miller, “Big data: New opportunities and new challenges [guest editors' introduction],” *Computer,* vol. 46, pp. 22–24, 2013.

P. P. Tallon, “Corporate governance of big data: Perspectives on value, risk, and cost,” *Computer,* vol. 46, pp. 32–38, 2013.

R. Bergelt, M. Vodel, and W. Hardt, “Energy efficient handling of big data in embedded, wireless sensor networks,” in *Sensors Applications Symposium (SAS), 2014 IEEE,* 2014, pp. 53–58.

P. Malik, “Governing big data: principles and practices,” IBM Journal of Research and Development, vol. 57, pp. 1: 1–1: 13, 2013.

C.-W. Tsai, C.-F. Lai, M.-C. Chiang, and L. T. Yang, “Data mining for internet of things: a survey,” Communications Surveys & Tutorials, IEEE, vol. 16, pp. 77–97, 2014.

R. M. Chang, R. J. Kauffman, and Y. Kwon, “Understanding the paradigm shift to computational social science in the presence of big data,” *Decision Support Systems,* vol. 63, pp. 67–80, 2014.

S. Fang, L. Da Xu, Y. Zhu, J. Ahati, H. Pei, J. Yan, *et al.*, “An integrated system for regional environmental monitoring and management based on internet of things,” *Industrial Informatics, IEEE Transactions on,* vol. 10, pp. 1596–1605, 2014.

D. J. Cook, “How smart is your home?,” Science (New York, NY), vol. 335, p. 1579, 2012.

L. Atzori, A. Iera, and G. Morabito, “The internet of things: A survey,” Computer networks, vol. 54, pp. 2787–2805, 2010.

A. Aijaz, “Cognitive Machine-to-Machine Communications for Internet-of-Things: A Protocol Stack Perspective,” *Internet of Things Journal, IEEE,* vol. 2, pp. 103–112, 2015.

G. Yang, L. Xie, M. Mantysalo, X. Zhou, Z. Pang, L. Da Xu, *et al.*, “A Health-IoT Platform Based on the Integration of Intelligent Packaging, Unobtrusive Bio-Sensor, and Intelligent Medicine Box,” *Industrial Informatics, IEEE Transactions on,* vol. 10, pp. 2180–2191, 2014.

S. Chen, H. Xu, D. Liu, B. Hu, and H. Wang, “A vision of IoT: Applications, challenges, and opportunities with china perspective,” *Internet of Things Journal, IEEE,* vol. 1, pp. 349–359, 2014.

Chapter 3
Wellness Protocol Development and Implementation

Abstract In the present research, the WSNs-based Wellness Protocol System for home monitoring is planned and realized two levels; hardware and software. At the hardware level, heterogeneous sensors are installed to get multi-activity and multi-event; these wireless sensor nodes are developed on Intel Galileo. The sensors, XBee module (as RF transceiver only) are connected and programmed to Intel Galileo board. Data is received through central coordinator node and collected into local home gateway computer server. The software logic has been developed for wellness protocol. The software module is subdivided into different levels, such as data logging, data extraction, and data storage. One of the important tasks of the task of the software module is to forecast the change in activity and correlate it with the wellness of the occupant in near time or real time (Ghayvat et al. in IEEE Sens J 15:7341–7348, 2015a). Intel Galileo-based intelligent monitoring sensing system have been designed that operates on the wellness protocol and uses the features of IoTs. The sensors are interfaced to Intel Galileo. Galileo board processes sensor data by two algorithms, one is packet encapsulation, and another is intelligent sampling and control. The packet encapsulation algorithm is common for all sensing nodes in a network. While the intelligent sampling and control algorithm is programmed individually according to sensors characteristics and application. In the present chapter, the details of sensing node development, device configuration, deployment, wireless data communication, storage, and analysis approach of heterogeneous sensor data fusion have been documented.

3.1 A Brief About Wellness Protocol System

The wellness protocol based smart home system is equipped with sensors to monitor the activity of an individual. Figure 3.1 shows the representation of the brief picture of smart home monitoring from node level to decision level. The daily routine of object usage, movement, and change in other ambient parameters is recorded into sensing units. The sensor activation data generated from daily routine is sent to coordinator-gateway system through wireless communication. The

H. Ghayvat and S.C. Mukhopadhyay, *Wellness Protocol for Smart Homes*,
Smart Sensors, Measurement and Instrumentation 24,
DOI 10.1007/978-3-319-52048-3_3

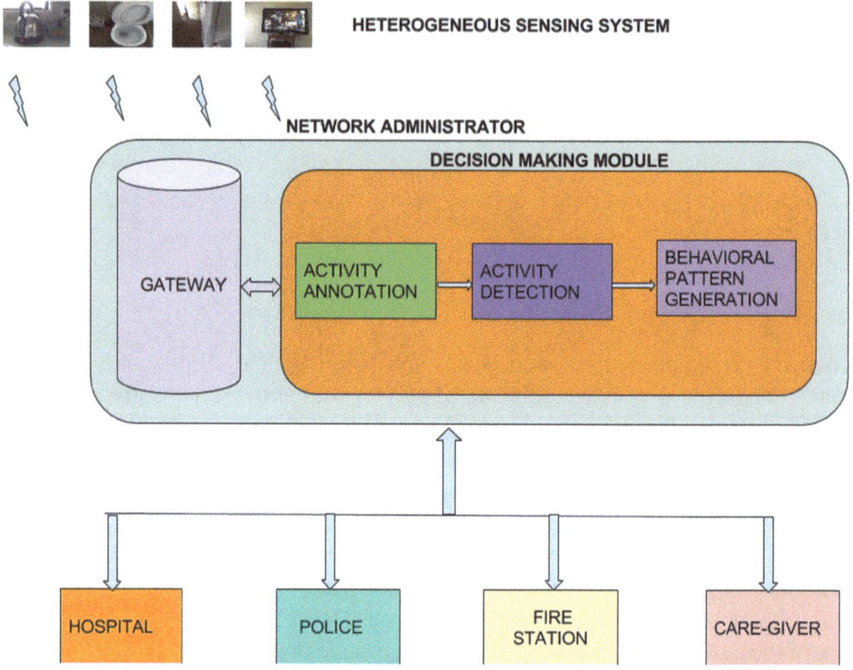

Fig. 3.1 Representation of wellness protocol based home system

network administrator block is the local home gateway server which collects, stores and perform the decision-making process. The decision process is done inside the decision-making module, first the annotation or labeling of data according to their sensor ID, time stamp and location, followed by activity recognition and wellness pattern generation. The generated information is sent to the caregiver, healthcare professionals, home security experts, and police as desired.

3.2 Wellness Approach to Protocol Development

The Wellness Protocol targets an event and priority-based data communication. It offers reasonable packet delivery metrics and large data handling. The protocol intends to cover complete smart home solution, starting from the sensor node to real-time analysis, data streaming, decision-making, and control. Figure 3.2 represents the architecture of wellness protocol based smart home solution; WSDA stands for Wellness Sensor Data Acquisition.

The project of smart home monitoring and control for AAL was initially started with ZigBee protocol but later the limitation of ZigBee triggered the demand of new protocol. The several header fields of the ZigBee stack are a huge overhead for

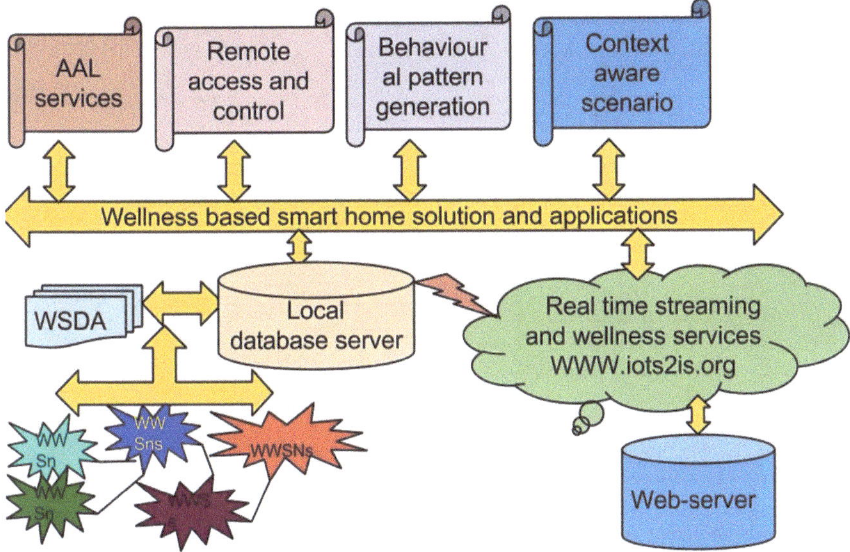

Fig. 3.2 Functional description of the developed smart home monitoring system

ambient assisted living. Foran example, ZigBee uses 64-bit headers for addressing; the encryption algorithm used is **AES** (**Advanced Encryption Standard**) with a **128b** key length (16 Bytes) (Kioumars and Tang 2011). The Fig. 3.3 shows the IEEE 802.15.5 ZigBee stack fields and auxiliary security header in detail. Whereas Fig. 3.4 represents the IEEE 802.15.5 ZigBee data payload field in the detail.

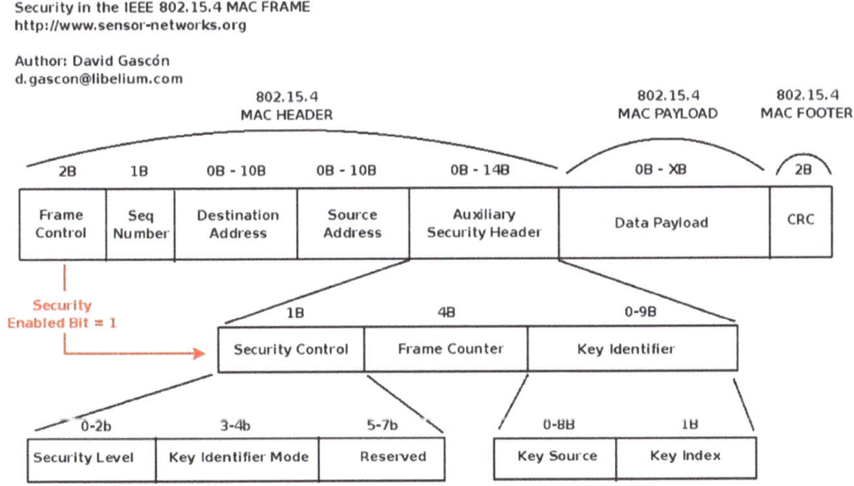

Fig. 3.3 Representation of IEEE 802.15.5. ZigBee stack fields and auxiliary security header in detail

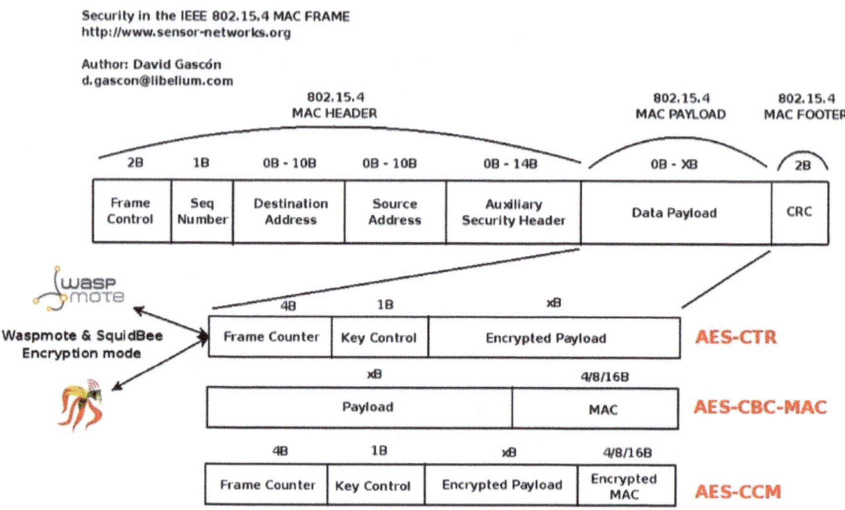

Fig. 3.4 Representation of IEEE 802.15.5. ZigBee stack fields and data payload in detail

These limitations of IEEE 802.15.4/ZigBee motivated to design wellness protocol. Wellness Protocol includes the basic role fields for successful data processing and delivery. These fields are as follows.

Packet Fields

Data Recovery

Start and end delimiter [16 bits]: To identify every new packet, data extraction logic needs the starting and ending signature.

Loss Detection

Frame count [4 bits]: This frame count is the counter to count the number of packets sent by an end node and received at coordinator-receiver. The 4 bits can count 16 data packets. This field is inevitable in case lost packet detection.

Routing

Source and Destination address [8 bits]: Single as well as multi-hop communication devices require the routing path that is identified according to source and destination address.

Error Detection

Checksum [8 bits]: The urban home environment is a medium with significant path losses caused by the interference of other devices and attenuation of different materials present in the vicinity. To detect the error in the received packet, checksum field has been introduced.

Context, Priority, and Event-based Information

Sensor type or event priority [4 bits]: The sensor data payload field indicates the type of sensor data. This field contains a unique number to represent the type of

Table 3.1 Priority of events in descending order

Serial number	Priority in descending order
0000	Panic button press
0001	Fall
0010	Smoke
0011	Medicine
0100	Toilet
0101	Food
0110	Sleep
0111	Personal hygiene
1000	Entertainment
1001	Relax
1010-1111	Reserved for future

event and its priority. Every unique number has a separate definition that decides its priority during analysis for decision-making. This decision-making offers the context-aware features. Few sensing events have the highest priority among others such as the fall detector has the peak priority. If the person under monitoring has fallen, the corresponding sensor data will be sent without any delay and other events will be queued. Table 3.1 represents the priority of different activities.

Security

- Encryption: One of the challenging issues for adaptation of smart home is privacy. People hesitate to be monitored due to the insecurity of data measured. The wellness dynamic key generation (WDKG) had been used to ensure the security of payload.

The proposed wellness packet encapsulation is a compact approach that takes care of packet size; event and priority based data processing. Figure 3.5 represents packet encapsulation at the end-device station with the event and priority based architecture; the CTR stands for the container.

The next sub-section describes the issues found while practical implementation of ZigBee for Smart Home and solutions offered by Wellness Protocol for the same.

3.2.1 Intelligent Sampling and Transmission Control Algorithm

One of the significant parameters for raw sensor data transmission is sampling frequency. The ZigBee protocol based XBee RF module is a configurable device and the configuration can be done through XCTU software (Digi 2016). This XCTU software is freeware software. ZigBee offers the periodic sampling options. This sampling rate ranges 0 Hexadecimal to FFFF Hexadecimal. With the sampling parameter set, the radio will periodically sample all of the I/O lines that

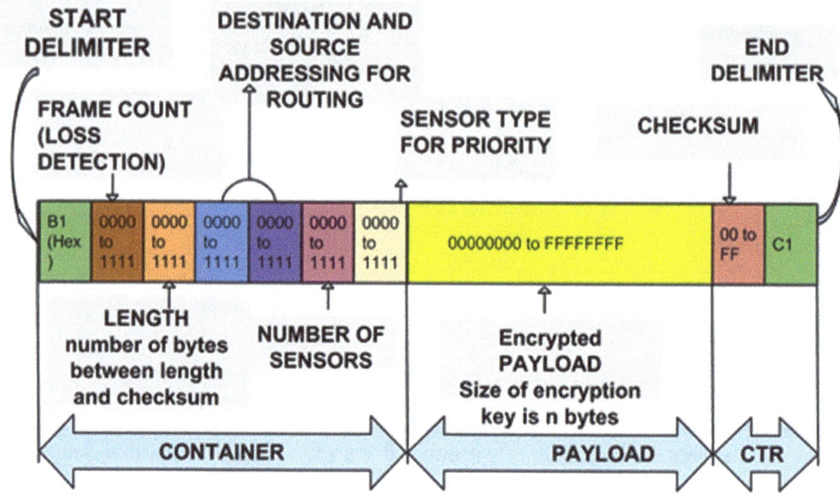

Fig. 3.5 Event-priority based packet creation for WSNs for smart home solution

are set as inputs and send the sample data to a destination XBee specified by DH and DL. Ultimately, the sampling rate decides the large data size which stores at the server. In ZigBee, the large data usually is controlled in two ways, which is shown by Figs. 3.6 and 3.7. Figure 3.6 represents the sampling rate selected for the sensor node. In this approach, the more the sampling rate lesser the number of the packets, but this caused loss of useful information. Figure 3.7 shows the cyclic sleep period, where RF device was kept on sleep and wake up. Cyclic sleep period had limited the number of packets up to a great extent but while its sleep the system missed some inevitable data packets.

Initially, the common packet encapsulation standard was selected for all sensors in the wellness approach. No intelligent algorithm to control data flows has been applied. The system recorded too much excess data resulting in a problem of storage. Data did not synchronize with activities and devices, so, there was a significant need to design a new algorithm according to sensor application, introducing event and priority. ZigBee protocol based XBee RF module in API, as well as AT mode of radio communication, offers the sampling frequency from milli-sec to minute. Let us assume that the sampling rate is f_1. After every f_1 time period corresponding to the frequency, the end-device delivers the data to coordinator-receiver regardless of activity and event. If any device has a high sampling rate, then it will send a higher amount of data to the receiver. This makes the data reception, storage, and analysis difficult. So; there is a need for control of transmission of data.

The smart sampling rate is essential to filter the excess, repetitive and unwanted data every time. Additionally, the algorithm must not lose and block the useful data packets. This algorithm is the entry module where raw sensor data comes into wellness protocol. During the initial phase of algorithm designing, it did not get

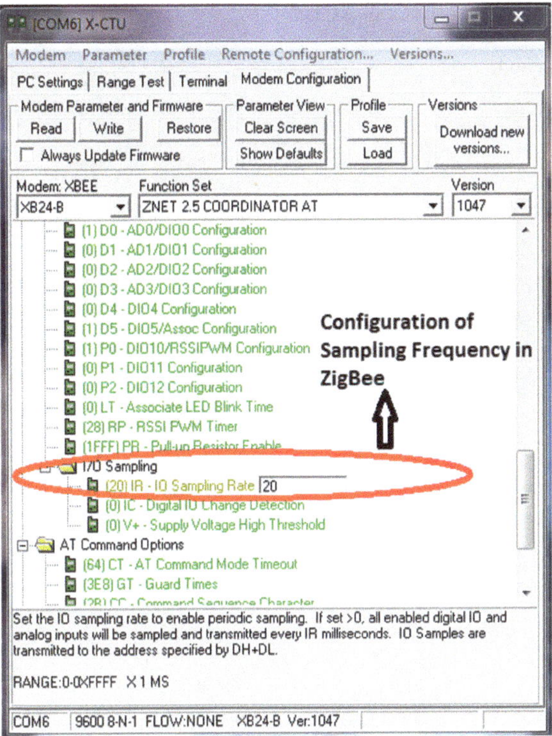

Fig. 3.6 Snapshot of XCTU to show the periodic sampling rate selection

reliable and appreciable performance. At times even after intelligent sampling system got too much unnecessary data or lost many packets blocked by algorithm logic. Upon examination, quite a few issues have been discovered, the dominant one being noisy sensor output voltage. There were some other issues as well, such as RF interference, noise power references, and accuracy of the sensor. The algorithm has been modified and upgraded to resolve these issues by introducing the advanced filter logics.

Figure 3.8 represents the intelligent sampling and control algorithm. When power is supplied to the intelligent wireless sensor node, the sensor generates the data and forwards it to processing and control unit. This control and the processing unit are configured and programmed according to intelligent sampling and control algorithm for the purpose of effective data transmission and flow control. The algorithm initiates with pin configuration and variables initialization for raw sensor data input. The ADC value of this data is generated and collected in the processing memory of Intel Galileo Board. Later on, this value is passed through advanced filter logic. These filter modules are good and have been specially designed to overcome the issues of excess data flow and blocking of useful data. After this filter check, the data is analyzed at two levels. First, when the desired event takes place, there is sufficient and inevitable information to be delivered. Second, the data generated by the sensor is avoidable, but it reaches time threshold value (time-out).

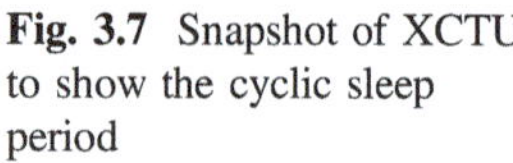

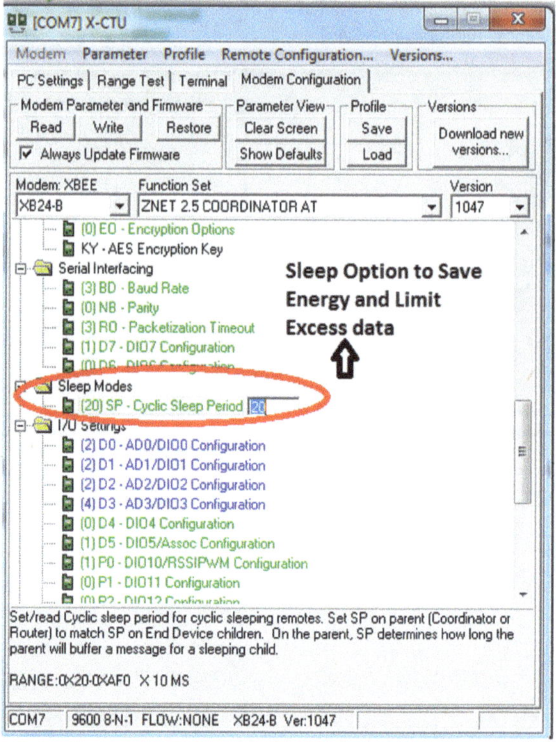

Fig. 3.7 Snapshot of XCTU to show the cyclic sleep period

The data is transmitted to the next stage of wellness model only if the sensor data passes any of the two levels. In the next phase, encryption algorithm takes the control for wellness packet generation. This is the combination of intelligent judgment and analysis for protocol generation.

The intelligent sampling and transmission control algorithm handles synchronization between sensor nodes and connected devices in the network. The time synchronization is a function of event and activity based data communication in the network. The network time synchronization decides the packet encapsulation and transmission without losing useful information.

Figure 3.9 shows the data collected from E & E sensing unit for wellness and ZigBee protocol. The data has been collected for three different sampling rate of ZigBee and compared with wellness for 1 h to 27 days. About 1200 kB data have been stored in the local home gateway server in 27 days for wellness based six E & E sensing nodes deployed in the smart home, while for the same duration the ZigBee has noted 687 kB (10 min sampling rate), 6865 kB (1 min sampling rate) and 411,936 kB (1 s sampling rate). The data recorded in ZigBee system was very high as compared to wellness system with sampling rate 1 min and 1 s, while 10 min are sampling rate caused by sampling.

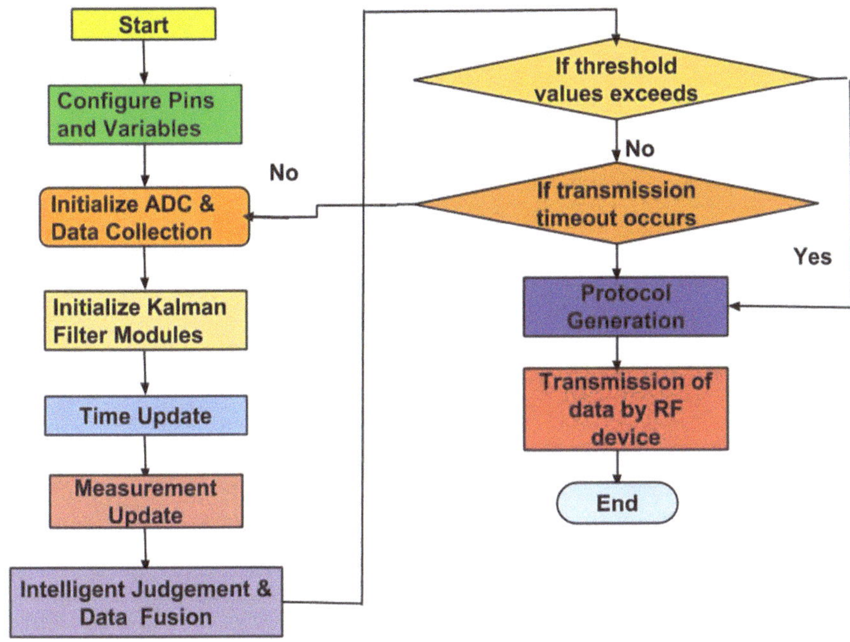

Fig. 3.8 Intelligent sampling and transmission control algorithm

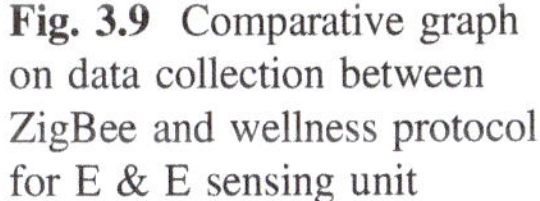
Fig. 3.9 Comparative graph on data collection between ZigBee and wellness protocol for E & E sensing unit

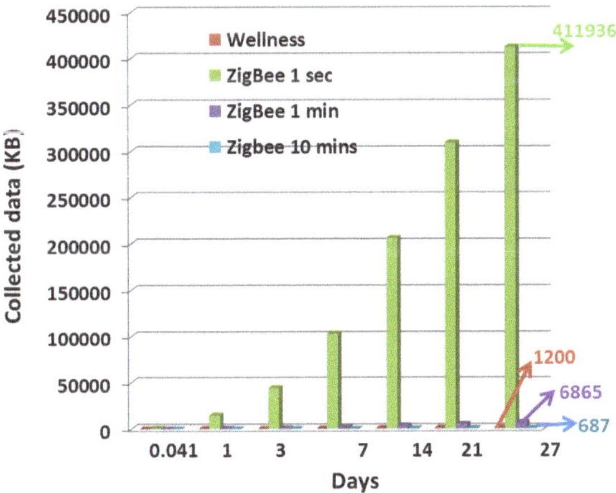

Figures 3.10 and 3.11 represent the comparative analysis between the wellness protocol and ZigBee protocol based smart home for two different houses. The maximum possible sample rates have been selected for ZigBee system in both the houses. Even with maximum possible sampling rate the data recorded in ZigBee system in both the houses is very high as compared to a wellness system. For subject-house one and subject-house two the ZigBee-based system recorded

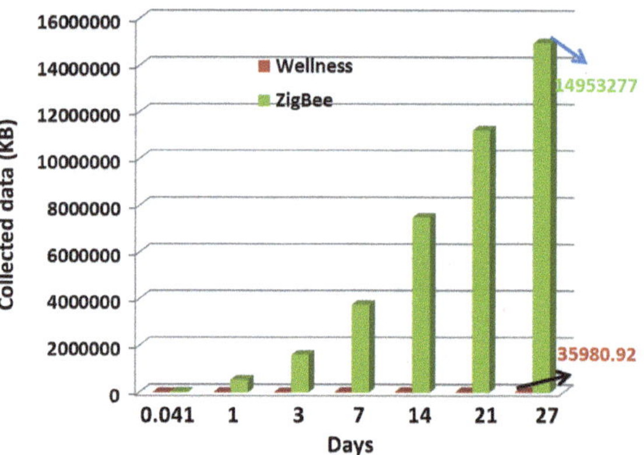

Fig. 3.10 Comparative graph on data collection of smart home system for House 1

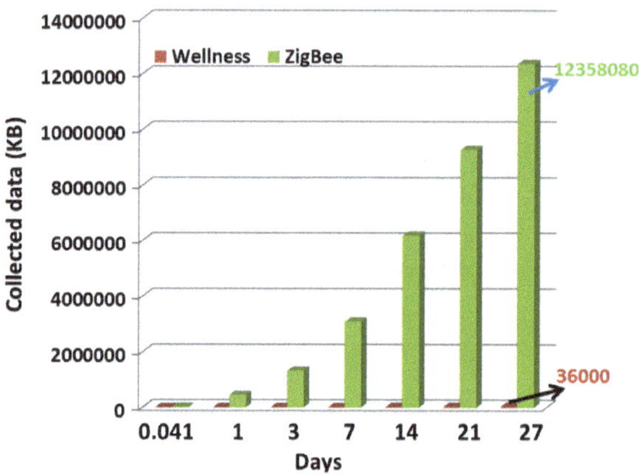

Fig. 3.11 Comparative graph on data collection of smart home system for House 2

14,953,277 kB and 12,358,080 kB respectively, while wellness system touched 35,980 kB for subject-house 1 and 36,000 kB for subject-house 2.

The amount of data reduced with wellness based system was significant; about 410% data reduction had been achieved with zero loss of useful information packets. The large data handling was achieved without compromising the quality of information.

3.2.2 *Interference Mitigation*

Interference is any unwanted signal which disturbs the desired signal. In the present context, the interference is associated with the indoor devices which operate in unlicensed ISM band [128]. Figure 3.12 shows the channel selection in ZigBee through XCTU. After ZigBee system installation, the system recorded distorted data. This interference occurred because of other wireless devices operating and

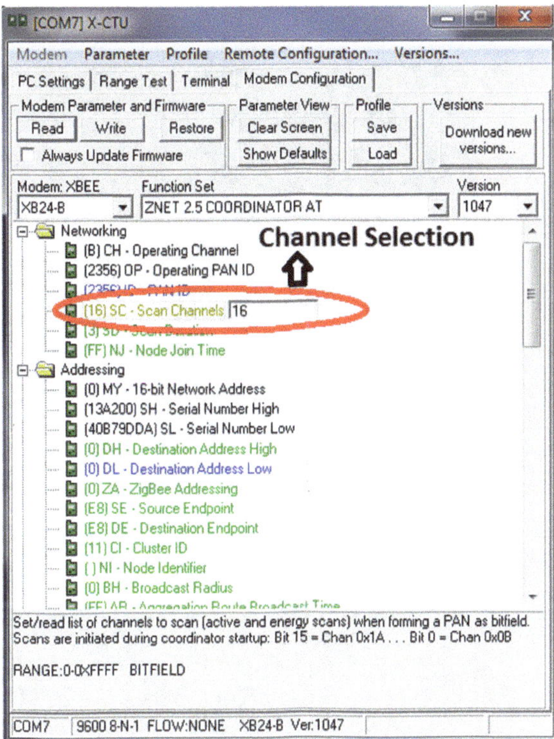

Fig. 3.12 XCTU snapshot for channel selection

overlapping the channel number 16 (2.430 GHz) of ZigBee as shown by Fig. 3.13. The best approach to coexist with other indoor interference sources was performing frequency spectrum analysis before selecting the channel. Metageek devise and frequency spectrum analyzer was used to analyze the interference sources and suggest possible mitigation approach (Metageek 2016; Ghayvat et al. 2016).

3.2.3 *Wellness Dynamic Key Generation for Security*

ZigBee had an issue with security overhead, so the best solution was to define a new security algorithm. Figure 3.14 shows the security enabling in ZigBee. Encryption provides secure communications in WSNs. Keys are the codes that algorithms use to encode and decode the information. These keys are shared between receiver and sender. There are three basic ways of keys arrangement schemes: the trusted server, the self-enforcing scheme, and the key pre-distribution scheme. In trusted server there is a server for key establishment between nodes. However, in the case of distributed heterogeneous sensor network, it is hard to build trusted server. The self-enforcing scheme uses asymmetric cryptography, such as a public key certificate. In home ambient, the sensor nodes are tiny and have limited computational resources, so the use of public key approach is not suggested. In key

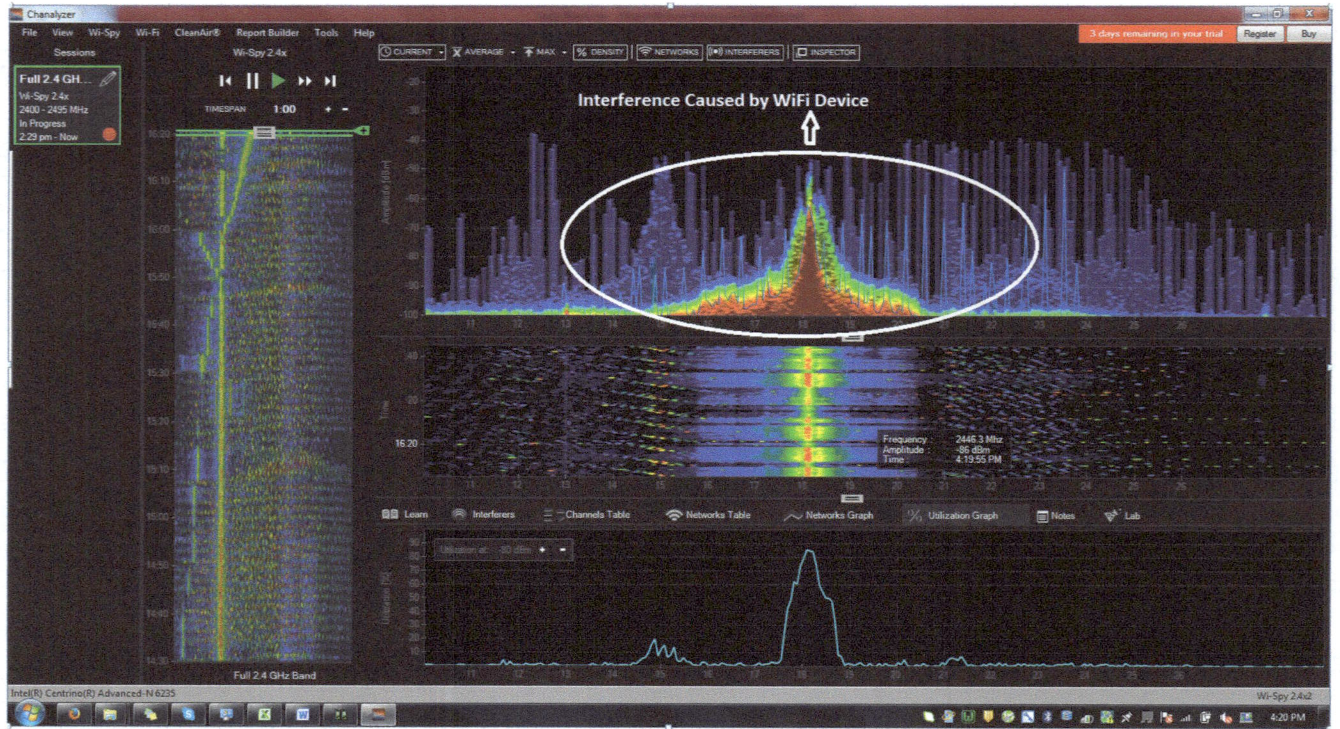

Fig. 3.13 Other household device functioning over the same channel where ZigBee device is operating

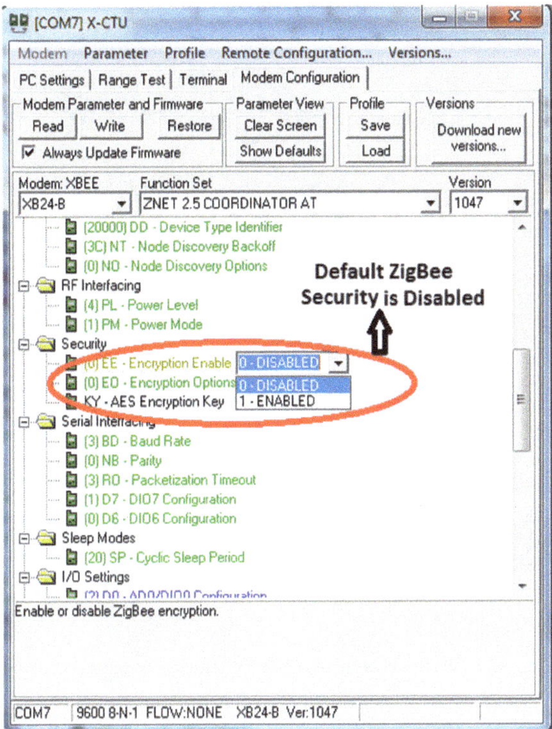

Fig. 3.14 XCTU snapshot to show the ZigBee encryption

pre-distribution schemes, keying materials are pre-loaded into sensor nodes before the deployment. This scheme offers the optimum performance with least complexity and computational cost for the small sensor network. The existing key distribution solutions require a large memory space in a WSN. There are some standard encryption algorithms available such as *AES, RSA, and Triple DES*. This complex algorithm takes significant processing and computational cost. For a smart home system where the number of sensors is not too much, the system can be implemented with a simple algorithm which demands less computational cost.

The purpose of the WDKG scheme is to resolve several security issues existing in the smart home system and also to overcome from the computational overhead and execution time. By considering the significance of real-time delay, caused by the authentication process and for supporting integrity and privacy of the real-time communicated data, the system uses the simple cryptographic primitives to ensure both the network security and real-time data security in wellness system. To acquire network security with least time consumption for authentication, the system has adopted the simple cryptographic primitives such as EXCLUSIVE-OR operations, which has caused reasonably less execution time and computational overhead as compared to any symmetric or asymmetric key standard. Figure 3.15 presents the security scheme designed in Wellness Protocol. The dynamic key generation takes place at two levels, at first level, the key is generated separately for each sensor

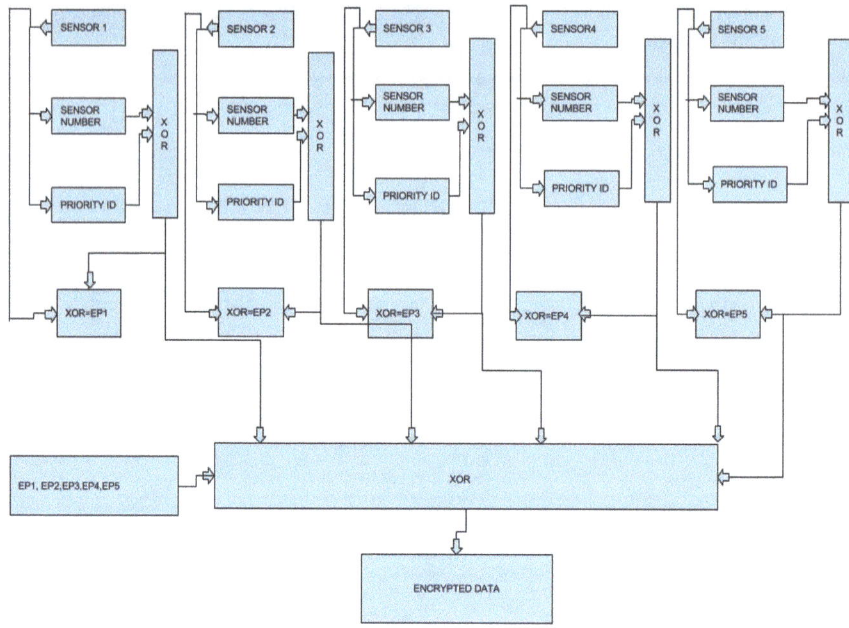

Fig. 3.15 Wellness dynamic key generation based security algorithm

through the xor operation between priority ID and sensor number field. At second level, the common key is generated through the xor of individual keys generated at level 1. Encryption of payload data also takes place at two levels, first by individual keys and second by a common key. The reverse operation is performed at the receiver by x-nor function.

The implemented security algorithm has been described by the equations from 3.1 to 3.8. Equation 3.1 measures the individual key for each sensor separately for level one. Equation 3.2 calculates the common key for level two. The individual sensor payload encryption is done by Eq. 3.3, and final stage encryption is done by Eq. 3.5. The reverse of this is performed by Eqs. 3.6–3.8.

Key Generation

Where “i” is a number of sensors, N is unique sensor number, P_r is a priority of sensor in the sensing system, K is individual encryption key for level one and Key is common encryption key for level two.

$$K_i = \sum_{i=1}^{n} N \text{ xor } P_r \tag{3.1}$$

$$\text{Key} = K_1 \text{ xor } K_2 \text{ xor } K_3 \text{ xor } K_4 \text{ xor } K_5 \tag{3.2}$$

Encryption process

Where “D_i” is each sensor data

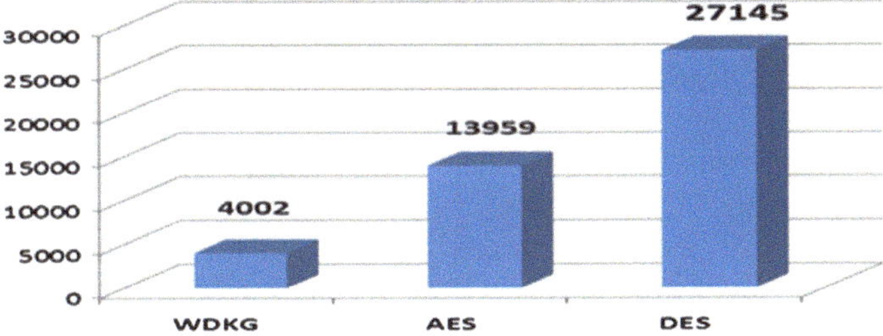

Fig. 3.16 Performance benchmark according to CPU cycles

$$E_i = D_i \text{ xor } K_i \tag{3.3}$$

$$P = \sum_{i=1}^{n} E_i \tag{3.4}$$

$$E_D = P \text{ xor Key} \tag{3.5}$$

Where "E_D" is encrypted data
Decryption process

$$P = E_D x - \text{nor } K \tag{3.6}$$

$$E_i = P/i \tag{3.7}$$

$$D_i = E_i x - \text{nor } K_i \tag{3.8}$$

The performance evaluation of WDKG can be done by comparing with existing encryption schemes. We can understand this better by the performance comparison. Figure 3.16 shows that AES takes 13,959 CPU cycles, DES takes 27,145 CPU cycles; WDKG takes only 4002 CPU cycles, and Fig. 3.17 shows the execution time of three schemes (Ghayvat et al. 2015c).

Fig. 3.17 Performance benchmark according to execution time

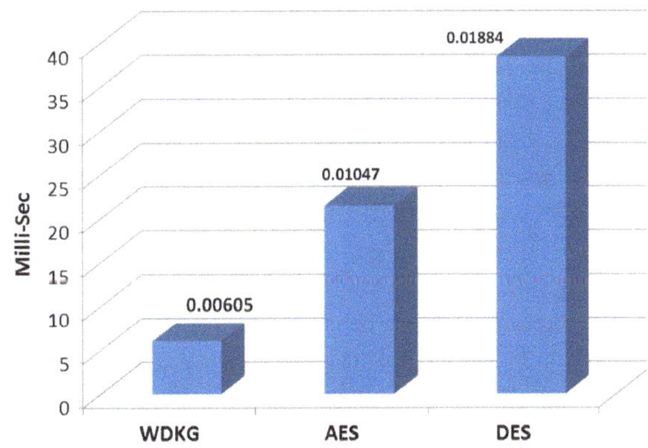

3.3 Wellness Sensing Units for Home Monitoring and Control

The system is implemented on two different levels, hardware, and software to offer integrated framework. The hardware system contains heterogeneous sensor network. This hardware network is further divided into four domains body sensor network (BSN), ambient sensor network (ASN), emergency sensor network (ESN) and other sensor network (OSN). The BSN is facilitated with fall detector. A wireless BSN provides long-term health monitoring of an occupant under physiological states without restraining their regular events. Though the individual who is under recovery would welcome to deploy the monitoring device over the body, the person who is fit and fine most of the time avoids this. The present wellness system only includes the physiological monitoring devices which are compact, and designed in such a way that it does not affect the daily routine of the individual. The second domain ASN comprises temperature sensing unit, movement sensing unit and forces sensing unit. For security and panic alert the system equipped with manual push button unit and fire detection in ESN. The last domain OSN offers the usage monitoring and control of electrical appliances through to electrical and electronic object sensing unit. OSN also includes the contact sensing unit. The software system is responsible for data collection, extraction, and storage into the server. Finally, server data is analyzed and processed by machine learning and data mining models to produce useful information for web-site and decision making.

The designed sensor network architecture is based on integrated framework. An integrated framework contains sensors which are grouped based on their operational characteristics, mobility and application in four different layers; BSN, ASN, ESN, and OSN, as presented in Table 3.2.

Taking comprehensive consideration of robustness, power consumption, practical deployment of wiring and decorate works, the below-listed sensors in the table have been selected. These heterogeneous sensors include PIR, temperature, force and electronic and electrical appliances monitoring unit, force sensor and manual push button unit. A quick overview over these sensing units is presented in Table 3.3 (Ghayvat et al. 2015b).

Prototype Design

With the prompt development of electronic products and the popularization of the network, Smart Home applications, IP cameras, smart plugs, the electric relay, has enabled many auxiliary control functions. However, in most cases, inhabitants have already decorated their home environment and are not prepared to change their original layout to do the wiring work. Thus, the proposed wellness system design has a heterogeneous wireless architecture, which can easily deploy over existing house layout, without doing any significant change. The wellness system is cost effective, flexible, real-time and assist-control. The Sensors integrated with Intel board as a front-end device to sample the living data from every living area can be

Table 3.2 The layer features and functions are presented

Layer	Functional characteristics	Event detection
BSN	1. Compact and mobile sensing unit which does not affect the comfort of users 2. The sensing device is deployed over body area 3. Battery powered sensing unit 4. Example: Impact sensor	– Fall detection
ASN	1. Sensors are deployed over fixed location in the home locations Example: Temperature sensing unit, movement sensing unit and force sensing unit 2. Mostly plugged with the home power supply. Except outdoor sensing unit which uses solar power, Example: Solar panel based outside temperature monitoring unit	– Temperature, pressure, and movement
ESN	1. It has same characteristics; it is mainly deployed for emergency and panic conditions which cause direct health and wealth risk 2. Example: Manual push button unit and fire detection	– Home security, fire, and any sudden health issue
OSN	1. It is also fixed sensing unit distributed in the different locations to monitor the usage of electronic appliances in the home environment. It also includes the contact sensor to monitor the use of an object such as a cabinet, desk 2. Example: Electrical and Electronic object sensing unit	– Use of toaster, fridge, television, computer, microwave, rice cooker and other appliances – Such as kitchen activities

easily attached to the wall, and work via a battery or home power supply. In this way, occupants only need to maintain the power supply after the installation of wellness system.

For the embedded system design there were many processing platforms to build sensing node, but the actual decision has done between Intel Galileo board and Raspberry Pi board. Both are open source, do-it-yourself (DIY) electronics hardware development boards featuring embedded processors. The table given below explains it more properly (Liu et al. 2015) (Table 3.4).

Intel Galileo, which contains Arduino Uno microcontroller along with Intel processor, makes processing and control fast and easy. The wellness protocol is designed and loaded on Intel baseboard. According to protocol logics and algorithms, the data is sent through RF module XBee. In Wellness, Protocol XBee is used as a standard transmitter and receiver. The raw data from the sensor is transmitted to the small Intel Galileo Microcomputer for intelligent sampling and

Table 3.3 Technical description and functioning of Sensing units

Sensor	Technical description and function
Force sensing unit	The flexiforce sensor is a Piezoresistive Sensor. A301 sensor has been used to get the amount of pressure applied to any object. When any pressure is forced on the sensor, the resistance of the sensor drops. Eventually, the output voltage rises. The range of resistance varies in force sensor specification found in different types of sensors. The pressure is ideally applied in the circular central portion of the sensor on both borders. The sensor is deployed underneath the objects to sleep and sit upon. The sensor is connected to a conditioning circuit with the 9 V power supply. It is analog as well as digital output based sensing unit
Contact sensing unit for domestic objects	For the purpose of household and everyday habit monitoring such as a self-grooming table, almirah, and office desk, wireless contact sensing units have been designed. The designed contact sensing unit connected to the household and everyday stuff to identify the frequency of usage, and these objects usage are monitored at local home gateway server by ON/OFF values. It is digital output based sensing unit
Temperature monitoring unit	To design temperature sensing unit, the LM35 IC is connected to the conditioning circuit. It is analog output based sensing unit. This sensing unit deployed in indoor and outdoor conditions. The outdoor sensing unit is deployed to obtain outside environmental temperature and compare it with indoor value
Movement monitoring unit	The passive infra-red (PIR) movement monitoring unit, is designed to identify the motion within the coverage range of the sensing system. This PIR sensing unit is compact, power-efficient, flexible and durable. These sensing units are also known as "IR motion detector." It operates on 5–12 V supply and the effective distance is less than 8 m, sufficient for one room in a smart home. These are binary mode sensors. It is digital output based sensing unit. The PIR sensing units are placed at the entry and exit of rooms, kitchen and bathroom
Electronics and electrical appliances monitoring unit	This power usage monitoring and control unit contains a transformer and other circuit modules. The transformer block comprises of voltage and current transformers. The voltage step-down transformer is used to transform the mains from 220 to 10 V signal, and the current transformer ASM010 is used to link the current in the line wire to the load through the current transformer circuit. For signal amplification, operational amplifier LM324 is applied associated with other components such as a rectifier, capacitors and gain resistors of specific parameters. The analog sensor signal output is supplied to the analog

(continued)

Table 3.3 (continued)

Sensor	Technical description and function
	channel of radio communication chip for wireless transmission. It is analog output based sensing unit
Manual push button unit	A novel approach is proposed and implemented to identify some of the activities, such as eating, taking medicine. It is difficult to detect these activities precisely by PIR or force sensors
Smoke-sensing unit	MQ-2 can be used for the detection of the Hydrogen Gas/Smoke/Alcohol/Butane/Methane/Propane. According to the Environmental Conditions for Gas Measurements, the using temperature range is from −20 to 50 °C. In an ideal volcanic environment, the temperatures are a bit higher during eruptions. So, this sensor is used for prove of the concept as it is cheap and affordable. A better sensor can be used in an ideal situation

Table 3.4 The reasons to select Intel board on the top of RPI

Metrics	Intel Galileo	Raspberry Pi (RPi)
Performance	Galileo runs the 400 MHz Pentium-class Quark	Raspberry Pi is usually clocked at 700 MHz, but since RPi performs fewer calculations per cycle, they are roughly equivalent in this aspect
Who is better for sensing application	Galileo is an outstanding choice if you have a project requiring sensors and value utility (requiring memory and processing power) and/or productivity or monitoring (Galileo has a real time clock)	Raspberry Pi is finest for media such as pictures or video
Cost	The Galileo board costs nearly twice as much as the Raspberry Pi Model B	There are some hidden costs with RPi, all that comes in the box is the board
Booting the boards	Galileo can boot from onboard memory	RPi requires formatting a card and copying the image before booting for the first time

control. This intelligent sampling and control analyses and processes the data before it passes to the event and priority based encapsulation stage.

This system addresses the issues of large data and delays significantly. The large data handling, delay, processing require the sorting and identification of inevitable data from other unwanted data at end-device (node level). In wellness scheme, an Intel Galileo board is required in the living areas. The sensing nodes integrate an XBee S-2, which enables the serial radio communication signal. The living or common room is typically located at the center of each house. So, Intel Galileo based coordinator receiver unit placed over there to get a reasonable distance from end-device. Then, other rooms like the bedroom, bathroom, kitchen, and study

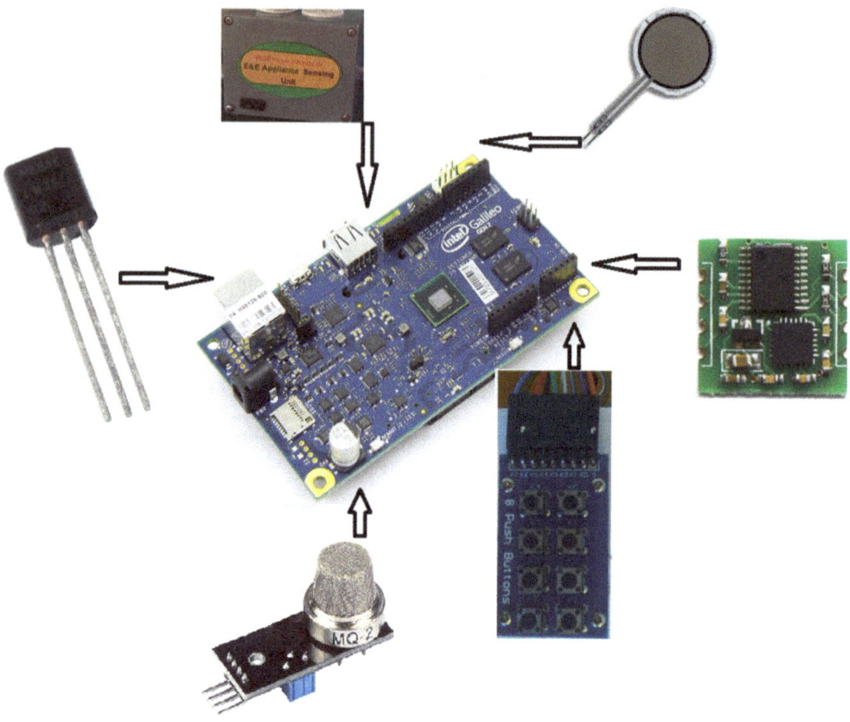

Fig. 3.18 Prototype of wireless sensing unit design

room where each requires one end device to gather the living information and environmental parameters. At first, HP Elitebook 8540 W PC workstation as a local home gateway server to collect, store and process the data has been installed. This server system enables the control of home appliances via a website. Moreover, the local home gateway server only needs to listen to the sampled data, process the data with a trained classifier and other data mining logics, and return the feedback control message to the destination. Figure 3.18 shows the prototype of wireless sensing unit design with Intel Galileo board (Ghayvat et al. 2014).

3.4 Wireless Topology, Network Formation for Smart Home System

RF module XBee can be programmed in any of the four possible ways by drivers as well as software. It usually functions in two different operating modes; the first is an application programming interface (API), and the other is application transparent (AT). XBee S-2 Mesh topology is the most preferred topology for the urban environment. Table 3.5 describes the possible device selections and features (Ghayvat et al. 2016).

Table 3.5 ZigBee device modes and their functions

XBee device type	Function
XBee coordinator (XC)	There is precisely one coordinator for every single network, and this is the device that takes the establishment network responsibility of the network. This coordinator is the transceiver, RF module that not only receives the data from respective sensor nodes connected to it, but also looks for remote configuration and fault detection of other associated sensor nodes in the network
XBee routers (XRs)	XBee routers (XRs) are applied to extend the network coverage area for wireless communication. The XRs transmit the packet of their neighbor to designated path
XBee end device (XED)	The XED cannot communicate data of other nodes; they are the last end of the network. They only pass their data towards the receiver. They are the source point of data
XBee end-device plus router (XEDR)	The XEDR node can convey its data and other neighbor's data as well in the network

Forming a Network

Digi XBee Series 2 is used as RF transceiver device in our smart and intelligent home monitoring system. The XC has the authority to select a channel, PAN ID (16-bit personal area network unique identification number that only belongs to particular XBee WSN), security policy, and stack profile for the network. Every time, to create communication XC starts with the proper channel and executes series of scans (energy scan or PAN scan) to realize any RF activity on various channels of XED, XR and XEDR. This energy scan can only notice the wireless sensor nodes (XED, XR & XEDR) which are registered with corresponding PAN ID to the XC.

3.5 Deployment of Heterogeneous Wireless Sensing Units in a Home

The present home monitoring system is continuously running since May 2015 without any significant complexity and maintenance; except minor precautions related to sensor physical damage and power supply. Figure 3.19 represents the image of the real smart home where monitoring system is running, and inhabitants are living their daily life. This house was built in 1938, so it is in an old home, but it has been converted to a smart home with the help of sensing technology. The layout representation of the smart home with sensor location and placement is shown in Fig. 3.20. This figure displays the genuine home environment where one that stays alone and has smart sensing placement to improve the wellness of individual i.e. Ambient Assisted Living (AAL). It shows different wireless sensing units for monitoring of environmental parameters and different domestic objects. In the present system, it contains heterogeneous sensors network; these heterogeneous

Fig. 3.19 Image of a real house where smart home monitoring and control system are installed

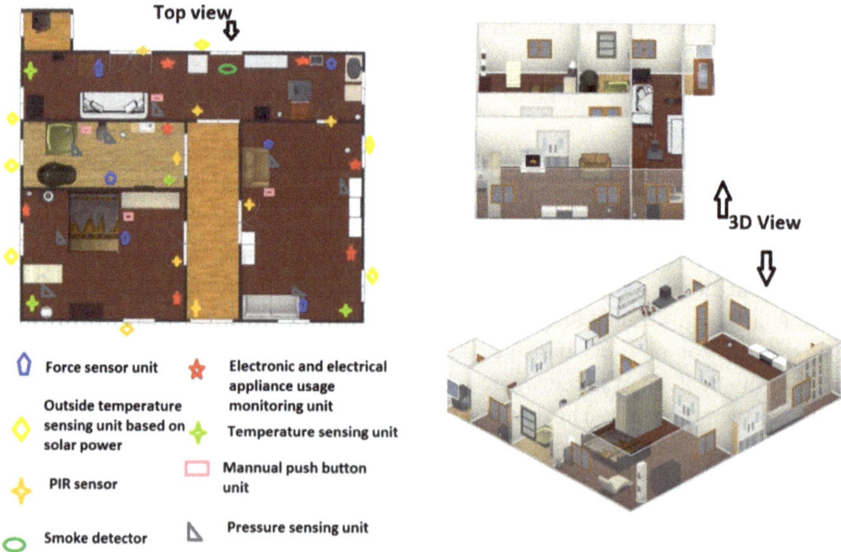

Fig. 3.20 Layout structure of the house with sensor deployment and household objects

Fig. 3.21 Outdoor sensing unit for outside temperature measurement

sensors are temperature, force, contact, smoke, PIR and electronic and electrical (E & E Appliances sensing) appliances monitoring unit. Figure 3.21 presents the outdoor temperature sensing unit to measure the outside ambient temperature. Figure 3.22 shows the utilization of E & E unit with different household appliances. This data is helpful to analyze the appliance usage pattern. Figure 3.23 shows the force sensing unit deployment to identify the occupancy. Figure 3.24 presents the deployment of PIR sensing unit to monitor the movement. While Fig. 3.25a shows the contact sensing unit connected to fridge and (b) shows the manual push button indicator (Ghayvat et al. 2015).

3.6 Healthcare

For the healthcare two sensing units are used in the present research work, one is fall detector, and another one is manual push button unit. While individual performs daily life activities, accidents such as a fall or an accident may occur in the home environment. As this might cause physical injuries, an instantaneous identification of an accident could alert humans for a fast reaction. Currently, an overabundance of inexpensive wireless motion sensing devices is available or can be assembled using off-the-shelf components. They can be used to collect motion data to monitor non-fall activities and to detect fall events. A framework has proposed to set up a wireless sensor network for collecting motion data and accelerometer sensor readings. A novel method has been introduced for semi-automatically extracting training examples from the motion data.

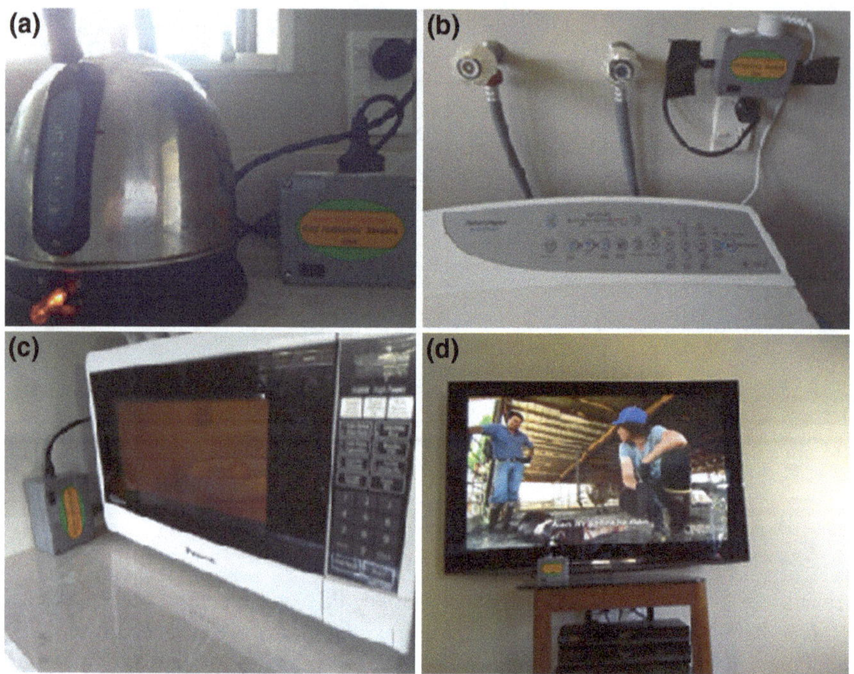

Fig. 3.22 E & E sensing unit, which is monitoring and controlling **a** Water kettle, **b** Washing machine, **c** Microwave and **d** Television

The accelerometer is an electromechanical sensing device that measures the static and dynamic forces of acceleration. Static force is a measurement of acceleration due to gravity whereas dynamic forces include the vibrations or any movement. Acceleration is the rate of variation of velocity measured. Accelerometers measure acceleration in either is one, two or three axes.

The MPU-60X0 Motion Processing Unit is the world's first motion processing solution with integrated 9-Axis sensor fusion using its field-proven and proprietary MotionFusion™ engine for handset and tablet applications, game controllers, motion pointer remote controls, and other consumer devices. The MPU-60X0 has an embedded tri-axis MEMS gyroscope, a 3-axis MEMS accelerometer, and a Digital Motion Processor™ (DMP™) hardware accelerator engine with an auxiliary I2C port that interfaces to 3rd party digital sensors such as magnetometers. Figure 3.26 shows the MPU-60X0 motion processing unit (Ghayvat et al. 2015d).

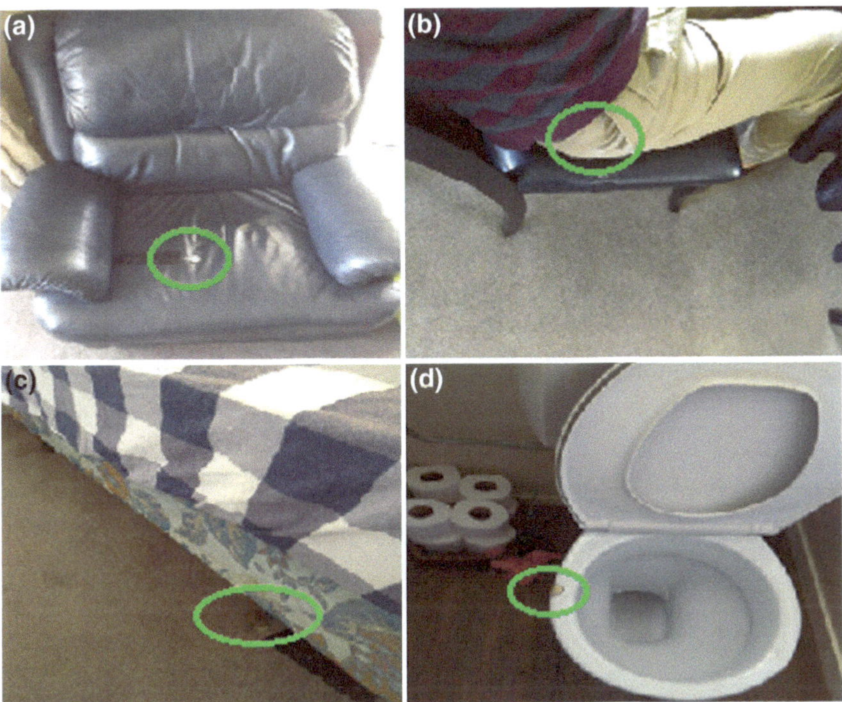

Fig. 3.23 Represents the force sensing unit deployment to monitor **a** Sofa, **b** Dining chair, **c** Bed and **d** Toilet seat

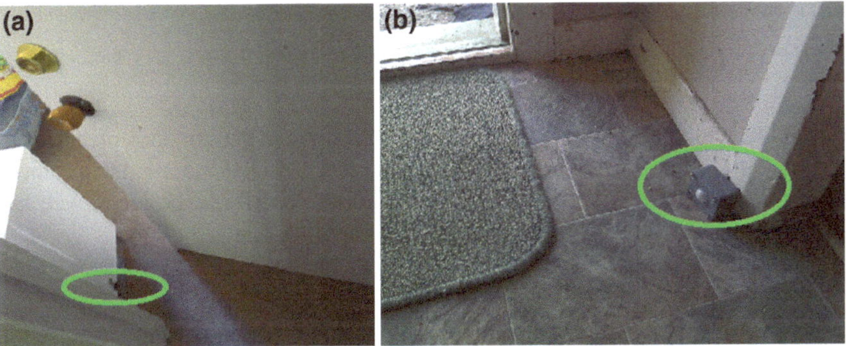

Fig. 3.24 Represents the PIR sensing unit deployment to monitor **a** Living room and **b** Entry door

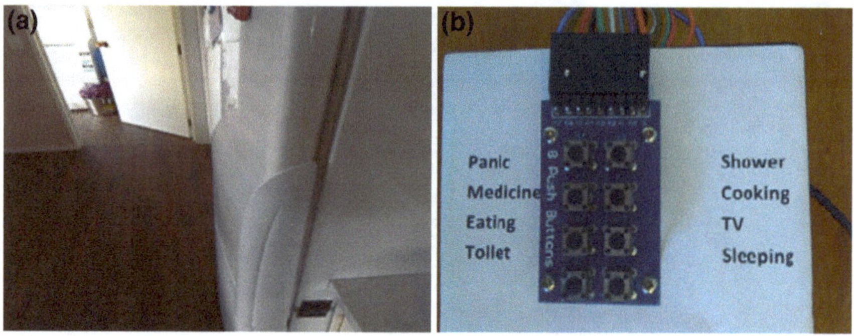

Fig. 3.25 **a** Contact sensing unit connected to fridge door and **b** Manual push button indicator

Fig. 3.26 The MPU-60X 0 motion processing units

3.6.1 *Software Description*

The software system, as depicted in Fig. 3.27 is designed and implemented to meet users' requirements and optimize performance. It consists of motion data protocol, fall detection algorithm, Angle calculation algorithm, wellness protocol, web server, database and Monitoring Device Application Software.

3.6.2 *Angle Calculation Algorithm*

The system collects gyroscope and accelerometer data while synchronizing data sampling at a user defined rate. The total dataset obtained by the MPU-60X0 includes 3-Axis gyroscope data, 3-Axis accelerometer data, and temperature data. The system calculated output to the system processor and got the roll, pitch, and yaw.

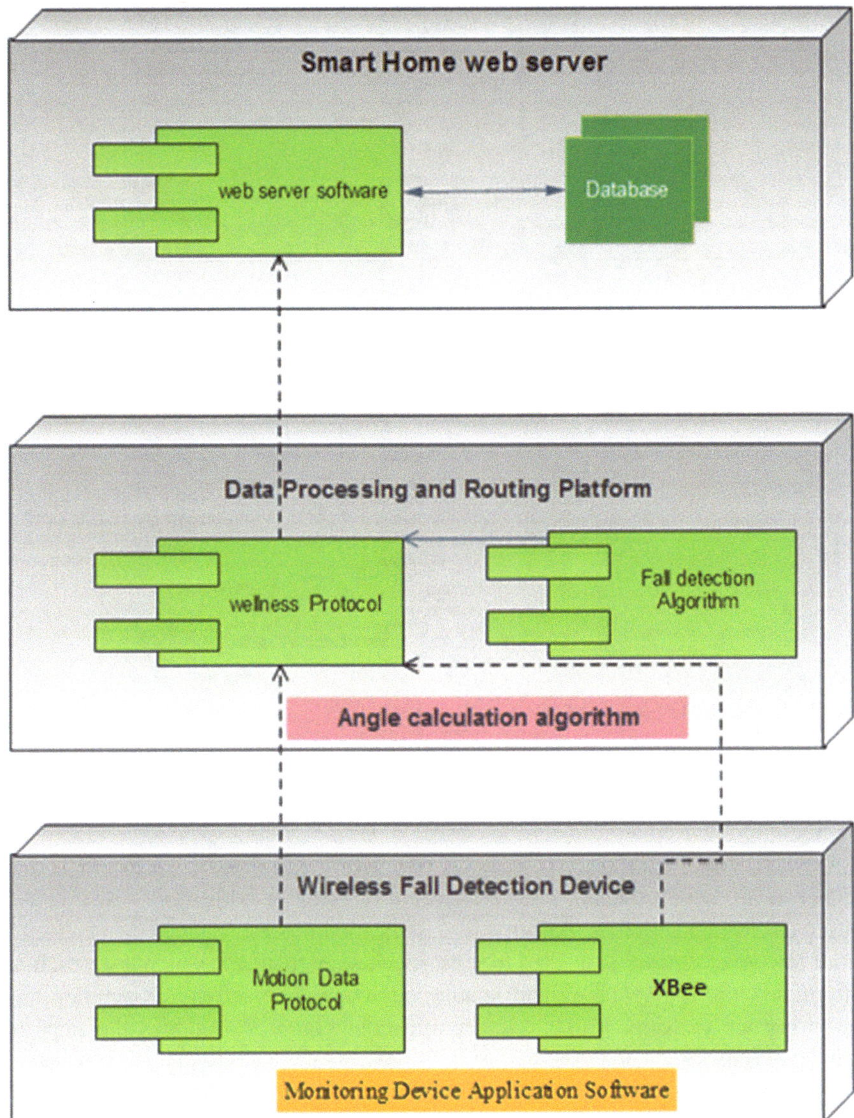

Fig. 3.27 Block diagram of software system

3.7 Desired Number of Wireless Sensing Units for AAL

The numbers of sensors in the present research are chosen on the basis of the lifestyle of an occupant. In this research, the sensors are deployed in such a way that it covers maximum possible home environment and usage.

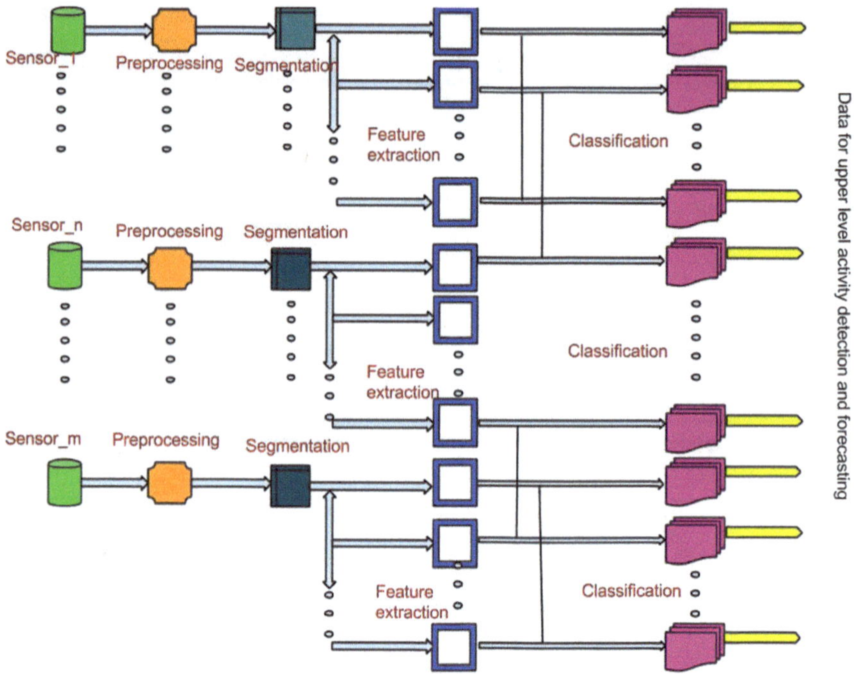

Fig. 3.28 The steps of low-level sensor data analysis

3.8 Data Extraction, Storage and User Interface

In a smart home, data analysis includes two levels: lower level sensory data and higher level activity data. Low-level sensory data classification converts the observations into activities. Before this classification process, the raw data collected from low-level sensors is needed to go through some data analysis phases such as pre-processing, segmentation and feature extraction. The activity recognition and discovery are performed after this classification. Figure 3.28 shows the steps of low-level sensor data analysis. Figure 3.29 presents the local home gateway server for wellness smart home system; whereas Fig. 3.30, the data comes from different heterogeneous sensors deployed in various locations inside the urban environment is collected into the local home gateway server through the coordinator. The wellness coordinator is connected to serial com port through the FTDI FT232R USB Com Port Driver. The data acquisition module is programmed and configured according to defined sensor data annotations and packet encapsulation fields for data extraction. The extracted data is sent to MYSQL data storage server (Bakar et al. 2016).

There is a deficiency of standard layout for storing data in the ambient assisted living environments research. This is restricting the prospect for researchers to compare and import-export the datasets of each other. These stored datasets are

Fig. 3.29 Local home gateway for Intel Galileo coordinator

processed further to extract the information; this information allows identifying the activities and behavioral patterns. The identified pattern and information can be used to diagnose the wellness of an occupant and provide the required support and care.

Indoor environment demands activity recognition of daily living from basic sensor data, and these raw data sets are composite and asymmetrical to translate into predefined scenarios. The decision-making information must be in a user-friendly

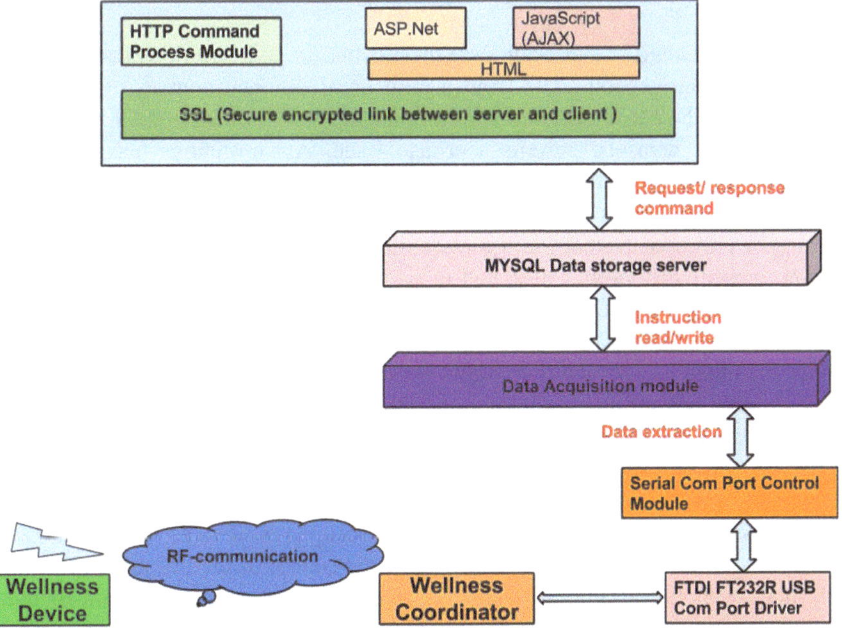

Fig. 3.30 Wellness protocol system architecture

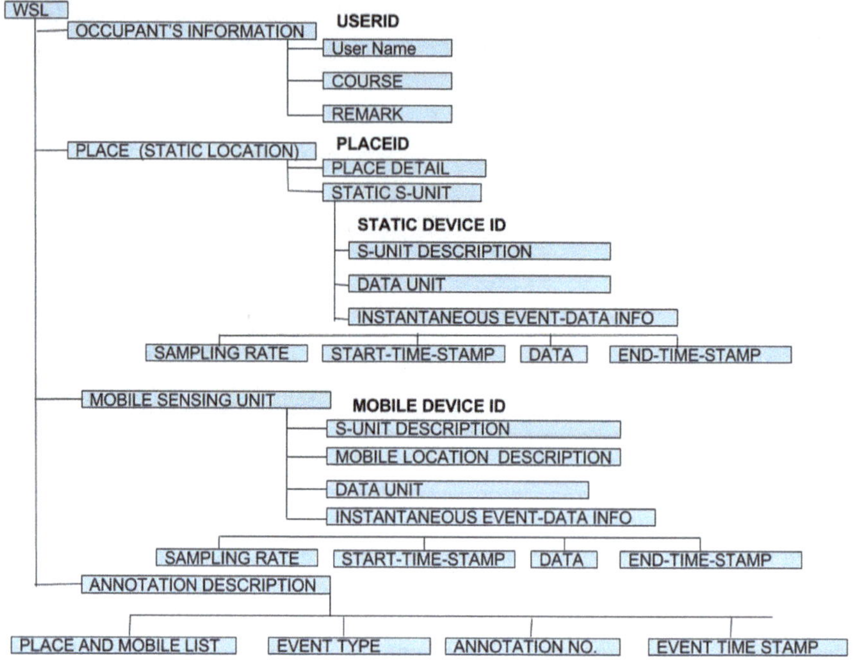

Fig. 3.31 Hierarchical representation of WellnessSF (Wellness standard format) tree diagrams for smart home system, (S-UNIT: SENSING UNIT)

format to the caregiver and heath care professionals. AAL solutions generate a big amount of heterogeneous data from a number of different sources of events via sensing unit. The current limitation of a standard format for annotation, storing and handling these datasets presents a critical issue. As resolutions become strictly reliant on the configuration of the sensing unit, the present research work attempts to resolve these issues through the creation of WellnessSF (wellness standard format), a proposed format for storing, annotations and sharing data.

The present system was designed to monitor the occupant into urban environment especially home conditions. The data was generated via fixed sensing units and body area network sensors as well. A location-based sensing unit is a device which is limited to a particular location, mostly integrated onto the defined ambient, such as force sensing unit. While on the other hand the mobile sensing unit is not constrained by a particular location. The mobile unit usually moves from one place to another, such as impact sensor. Figure 3.31 shows the hierarchical representation of WellnessSF tree diagrams (Khan et al. 2016). It shows the different sections of smart home data storage which includes occupant's details, the location of deployment, static and mobile sensing unit. Description of the event with real time information of time stamp and data is presented. The more description of these fields is given by tabular representation in Table 3.6. Figure 3.32 presents the architecture of home monitoring website.

Table 3.6 More description of user, location unit, mobile unit and annotation of tree diagram of WSL (Wellness Standard format)

Title	Explanation	Mandatory or Optional	Data type	Example
OCCUPANT is INFORMATION	Current wellness status with history.	Mandatory	Nil	Nil
USER ID	A unique identification number for an inhabitant	Mandatory	String	USER121
USERNAME	Occupant's name	Optional	String	Basant
COURSE	It shows the monitoring or care priority	Mandatory	String	Occupant USER121 needs the attention for turning on the room heater as temperature goes below 15 °C
REMARK	Other comments	Optional	String	Does not use back door
STATIC LOCATION	For ambient sensing unit fixed at one particular location	Mandatory	Nil	Nil
PLACE ID	A unique number to identify deployment area	Mandatory	String	Pl1
PLACE DETAIL	The common description of deployment location	Mandatory	String	Kitchen
STATIC S-UNIT	Location based static sensing unit details	Mandatory	Nil	Nil
STATIC DEVICE ID	A unique sensing unit identification number	Mandatory	String	S11
S-UNIT DESCRIPTION	Basic overview of the sensing unit	Mandatory	String	Electricity usage
DATA UNIT	Measured data unit	Mandatory	String	Watt
INSTANTANEOUS EVENT-DATA INFO	It shows runtime data packet details	Mandatory	Nil	Nil
SAMPLING RATE	Number of packet in defined time	Mandatory	Float	20
START-TIME-STAMP		Mandatory	Date and time	2016/02/10 17:21:08
DATA	Sensed event value	Mandatory	String	264
END-TIME-STAMP		Mandatory	Date and time	2016/02/10 18:12:32

(continued)

Table 3.6 (continued)

Title	Explanation	Mandatory or Optional	Data type	Example
MOBILE SENSING UNIT	The body area sensor or mobile sensing units which are movable			
MOBILE SENSING UNIT ID	A unique identification number	Mandatory	String	M121
S-UNIT DESCRIPTION	General detail of the sensing unit	Mandatory	String	Unit monitors the fall detection of a person
MOBILE LOCATION DESCRIPTION	General detail of mobile unit's deployment	Mandatory	String	Fall detection unit below neck
DATA UNIT	Measured data unit	Mandatory	String	Angle (degree)
INSTANTANEOUS EVENT-DATA INFO	It shows runtime data packet details	Mandatory	Nil	Nil
SAMPLING RATE	Number of packet in defined time	Mandatory	Float	26
START TIME STAMP		Mandatory	Date and time	2016/01/10 12:21:22
DATA	Sensed event value	Mandatory	String	221
END-TIME-STAMP		Mandatory	Date and time	2016/01/11 18:12:32
ANNOTATION DESCRIPTION				
PLACE AND MOBILE LIST	List of deployment locations and sensing units belongs to particular event	Mandatory	String	Temperature Sensing unit (S124) Force Sensing unit (S125)
EVENT TYPE	List all the events	Mandatory	String	Cooking
ANNOTATION NO	Unique identification number for any event or device or location.	Mandatory	String	A123
EVENT TIMESTAMP	Start and end of event	Mandatory	String	2016/02/10 13:12:43 to 2016/02/14 14:31:32

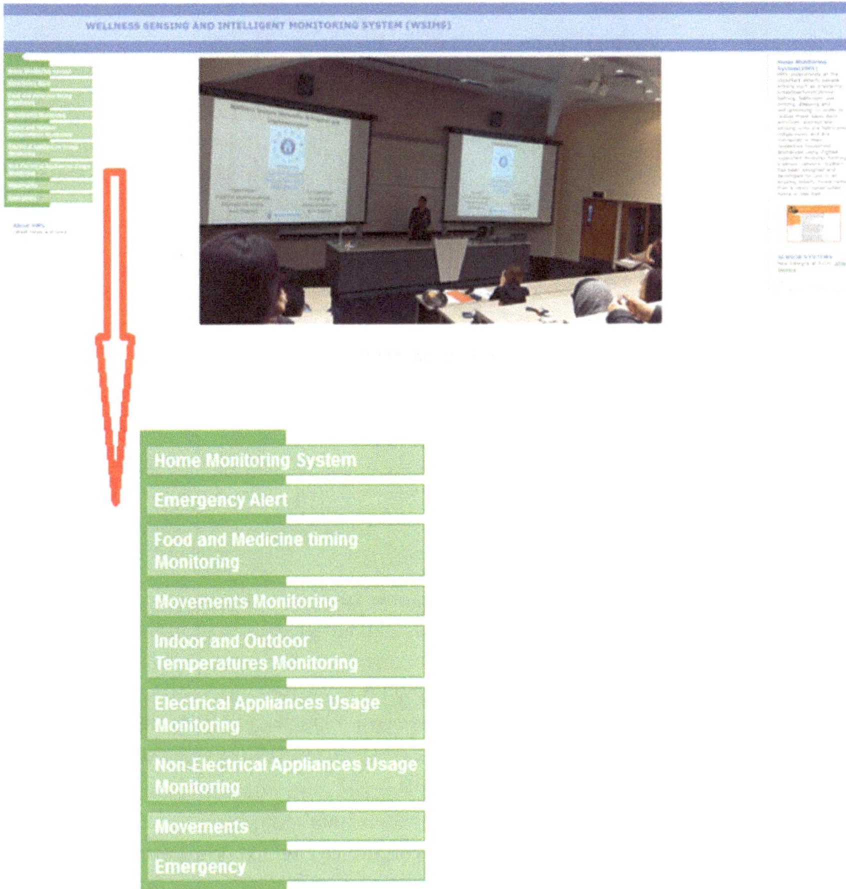

Fig. 3.32 Snapshot of home monitoring website which presents the sub-sections of monitoring

3.9 Evaluation of Wellness Protocol Data Reliability

To validate and evaluate the packet reliability of Wellness Protocol, a systematic and analytic research was carried out, and series of experiments were done in the smart home. Performance metrics in this wellness based smart home are assessed as an interdependent function of transmitter-receiver distance, type of building material placed between the transmitter(Tx)-receiver(Rx), type of hopping between wireless sensor nodes, the interference by other ISM frequency spectrum, sampling rate and a number of nodes. The dependent variables measured were packet delivery ratio, packet latency. The current section presents dependent variables for performance evaluation.

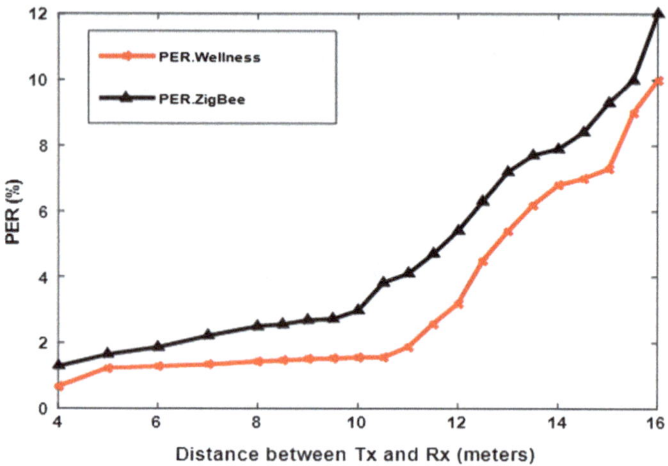

Fig. 3.33 Packet error rate of wellness protocol versus ZigBee with change in spacing between Tx and Rx

Figure 3.33 shows the packet error rate (PER) as a function of distance for Wellness Protocol based smart home and ZigBee-based smart home, under distributed hop. It was recorded that the packet error rate increased with the increase in distance significantly after 10 m distance for both ZigBee and wellness protocol, it was boosted up exponentially. Though the error increased in both the cases it got worse in ZigBee as compared to Wellness, at a distance 10 m the Wellness noted about 1.15% PER whereas ZigBee 2.28% PER and for the spacing 16 m the Wellness reached about 9.80% whereas Zigbee touched 12%.

Figure 3.34 shows the packet delivery ratio (PDR) of Wellness and ZigBee protocol as a function of distance, under multi-hop network setup. It was seen that the PDR remains constant till 8 m spacing but after that, it collapses in both the protocols. For the spacing 8 m, Wellness has recorded about 99.70%, whereas ZigBee about 99%. In both the cases beyond the 8 m distance PDR reduced, but in the case of ZigBee, it got worse with distance. For the spacing of 12 and 14 m, Wellness recorded about 50.21 and 16.63% respectively, while ZigBee touched 40.32 and 12.76% respectively.

Figure 3.35 shows the Packet loss rate (PLR) for ZigBee versus Wellness for varying spacing between transmitter and receiver. With the increase in spacing between Tx-Rx the PLR increased but it began to shoot up after the distance 8 m, and it became more inferior for ZigBee as compared to Wellness. At 8 m, the Wellness noted about 0.65% whereas ZigBee 1.38%. As spacing increased the PER increased exponentially. For the distance 12 and 14 m the Wellness reached about 61.35 and 79.92% respectively, while for the same ZigBee has touched 80.32 and 97.74% respectively. Figure 3.36 presents packet success rate (PSR) of ZigBee versus Wellness as a function of the distance between Tx-Rx. With the increase in spacing the PSR decreased and beyond the spacing 12 m, it falls very quickly.

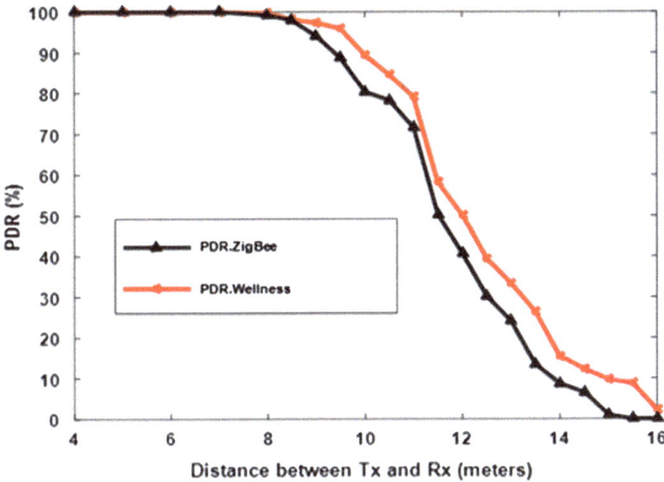

Fig. 3.34 Packet delivery ratio of wellness protocol versus ZigBee with change in spacing between Tx and Rx

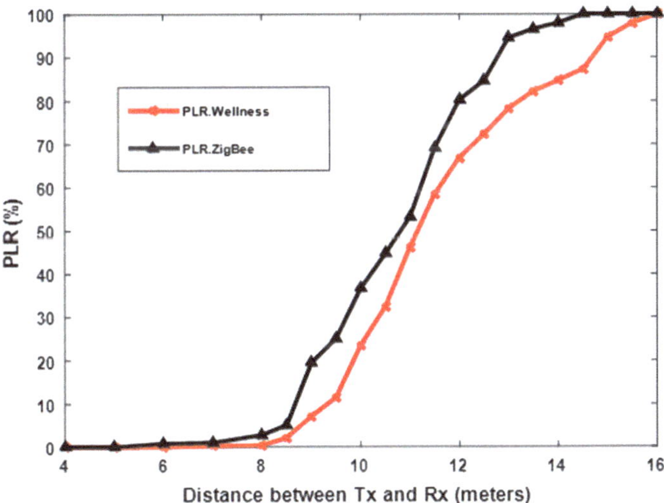

Fig. 3.35 Packet loss rate of wellness protocol versus ZigBee with change in spacing between Tx and Rx

Though in both Wellness and ZigBee the PSR fell with an increase in spacing it got poorer in ZigBee as compared to Wellness. For the spacing 8, 12 and 14 m the Wellness recorded about 98.90, 98.55, and 95.76% respectively, while for the same ZigBee s noted 98.70, 97.50, and 94.78% respectively.

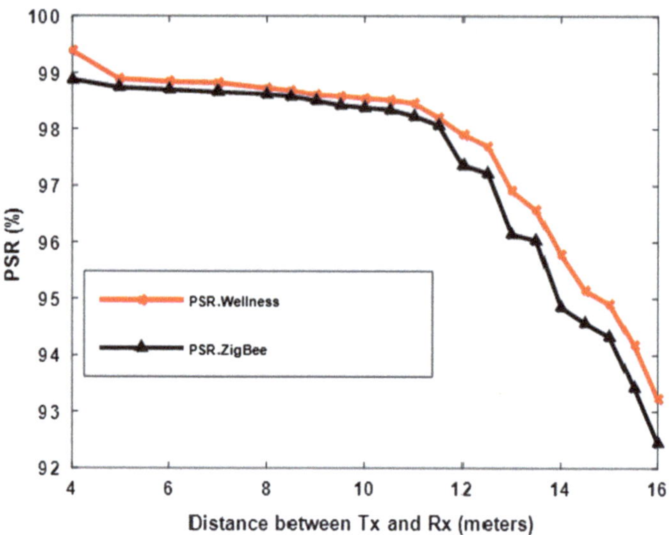

Fig. 3.36 Packet success rate of wellness protocol versus ZigBee with change in spacing between Tx and Rx

Figure 3.37 shows the average worst case delay on receiving data as a function of the distance between transmitter and receiver respectively. It was difficult to calculate the average delay of received packets, so, 10,000 samples for each spacing arrangement were observed and specially recorded the delay of the most delayed

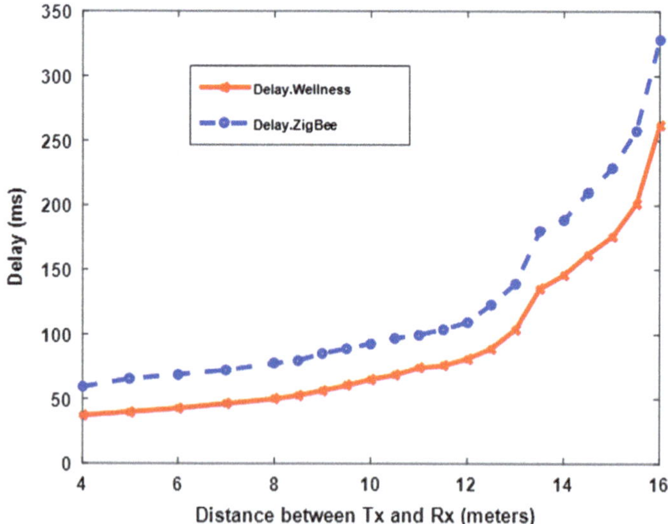

Fig. 3.37 Average delay of wellness protocol versus ZigBee with change in spacing between Tx and Rx

packet. It was recorded that average worst packet delay increases with distance in both the protocol ZigBee and Wellness. The amount of delay rises sharply after the distance 12 m; Wellness noted about 62 ms while ZigBee 91 ms, though the delay was in ms it was good enough to convert the real time monitoring system to near real-time monitoring system.

3.10 Software Required for Wellness System

The Wellness Home monitoring system consists of the following programs.

- A Real-time heterogeneous sensor data fusion application coded using C#. Visual Studio C# software 2010 used for it.
- MySql queries for the management of the near real-time sensor data storage in the database server. WAMP Server based MySql server is used.
- PHP scripts, JGraph scripts and Ajax programming codes for near real-time Sensor information displays on the web application.
- HTML is used to make web pages for the Internet website.
- Database replication procedures use MySQL scripts for connecting multiple home monitoring systems through the internet and Open-Virtual Private Network software.

3.11 Comparison of Wellness Over ZigBee

Comparison of ZigBee protocol and customized Wellness protocol can be better understood as follows.

Benefit of wellness packet encapsulation:

- Sleeping Routers and Battery life: Only end devices can sleep in ZigBee protocol whereas in Wellness all nodes can go to sleep. In Wellness system, Master as well as Coordinator has the responsibility of synchronization sleep and wake up to avoid the loss of data.
- Packet size: ZigBee (XBee S-2 API Mode) needs 21 bytes size to a single sensor for packet encapsulation, in comparison the Wellness Protocol requires 10 bytes.
- Latency: The processing delay is less in Wellness protocol as compared to ZigBee Protocol.
- Complexity related to data extraction from packet: It is significantly less in Wellness protocol as compared to ZigBee.
- Real-time operation: Wellness performs better than ZigBee for real-time web streaming of sensor information.

- Large data handling and storage: ZigBee is the generalized protocol for short-range monitoring that does not look after the large data handling and storage. The Wellness has focused on resolving the large data solution.
- Event and Priority based solution: The data packets usually get delayed due to the contention in the network and processing. In the wellness system, the data packets are transmitted and processed according to their rank (priority). If an event packet of health risk triggers, that packet would be sent and processed first.
- Packet reliability parameters for quality of service: Wellness has recorded notable improvement in packet reliability as compared to the poor reliability of ZigBee.

3.12 Conclusion

This chapter presents the development of Wellness Protocol for smart home monitoring. The design and implementation of heterogeneous WSNs have been described in detail. The designing of wireless intelligent sensing node, building the smart logic to overcome the issues and limitation of ZigBee have been achieved.

All the evaluation measurements are compared to ZigBee protocol. The large data handling is done by intelligent sampling algorithm which has offered the event and priority-based transmission and control; it has reduced data about 410%. The wellness dynamic key generation based security algorithm has recorded significant computational resources reduction; it uses 3.5 times less computational resources. The Wellness system performance has been measured for the data reliability parameters. Wellness system has noted an improvement of 3.46% for PDR, 1.58% for PLR, 1.84% for PER and 3.15 for PSR for the spacing 10 m. Whereas 32% delay reduction has been attained.

The Wellness Protocol is developed as an integrated platform which enables software and hardware to monitor the ADLs of an occupant. The following chapters are based on the application of the present system.

References

Ghayvat, H., Mukhopadhyay, S., and Gui, X.: 'Issues and mitigation of interference, attenuation and direction of arrival in IEEE 802.15. 4/ZigBee to wireless sensors and networks based smart building', Elsevier: *Measurement,* 2016, 86, pp. 209–226.

Ghayvat, H., Mukhopadhyay, S., Liu, J., and Gui, X.: 'Wellness Sensors Networks: A Proposal and Implementation for Smart Home to Assisted Living' *IEEE Sensors Journal,* 2015a, Volume: 15, Issue: 12, pp. 7341–7348.

A. H. Kioumars and L. Tang, "ATmega and XBee-based wireless sensing," in *Automation, Robotics and Applications (ICARA), 2011 5th International Conference on*, Wellington, 6–8 Dec, 2011, pp. 351–356.

Digi, (2016). *Digi XBee XCTU.* Retrieved Nov 19, 2016, from https://www.digi.com/products/xbee-rf-solutions/xctu-software/xctu.

Metageek, (2016). *Metageek device.* Retrieved Nov 19, 2016, from http://www.metageek.com/index-2.html?utm_expid=190328-155.9nhw_yWyRJK34laAPe_qUw.1&utm_referrer=https%3A%2F%2Fwww.google.co.nz.

H. Ghayvat, S. Mukhopadhyay, X. Gui, and N. Suryadevara, "WSN-and IOT-Based Smart Homes and Their Extension to Smart Buildings," MDPI: *Sensors,* vol. 15, pp. 10350–10379, 2015b.

Ghayvat, H., Liu, J., Alahi, M., Mukhopadhyay, S., and Gui, X.: 'Internet of Things for smart homes and buildings: Opportunities and Challenges', *Australian Journal of Telecommunications and the Digital Economy*, 2015c, 3, (4), pp. 33–47.

H. Ghayvat, A. Nag, N. Suryadevara, S. Mukhopadhyay, X. Gui, and J. Liu, "SHARING RESEARCH EXPERIENCES OF WSN BASED SMART HOME," *International Journal on Smart Sensing & Intelligent Systems,* vol. 7, 2014, pp. 1997–2013.

M. Khan, S. Din, S. Jabbar, M. Gohar, H. Ghayvat, and S. Mukhopadhyay, "Context-aware low power intelligent SmartHome based on the Internet of things," Elsevier: *Computers & Electrical Engineering,* ISSN No. 0045-79062016, pp. 1–15 (Early access).

U. Bakar, H. Ghayvat, F. Hasan, and S. Mukhopadhyay, "Activity and anomaly detection in smart home: A survey," *in Next Generation Sensors and Systems,* ed: Springer, 2016, pp. 191–220.

H. Ghayvat, S. C. Mukhopadhyay, and X. Gui, "Sensing Technologies for Intelligent Environments: A Review," in *Intelligent Environmental Sensing,* ed: Springer, 2015d, pp. 1–31.

J. Liu H. Ghayvat, and S. Mukhopadhyay, "Introducing Intel Galileo as a development platform of smart sensor: Evolution, opportunities and challenges," *in Industrial Electronics and Applications (ICIEA),* 2015 IEEE 10th Conference on, Auckland NZ, 15–17 June 2015, pp. 1797–1802.

Chapter 4
Issues and Mitigation of WSNs-Based Smart Building System

Abstract The functional characteristics and performance metrics of Institute of Electrical and Electronic Engineering (IEEE) 802.15.4 standard make it the only effective option in short-range environmental control and monitoring applications. The composite sensor and actuator nodes based on the wireless technology are placed in the building environment. Wireless technology deployed in building ambiance endures from the interference of various communication protocols operating in the same unlicensed, unregulated Industrial Scientific Medical (ISM) band, apart from the attenuation loss. A smart home designer could not omit these factors in the smart building ambiance because the adverse effects of these issues on the system performance are substantial. Most of the researchers have reported on adverse effects, but not with the aspects of a smart building. The current research reports on the detailed realistic-experimental analysis and mitigation for different types of interference, the attenuation losses, and direction of arrival associated with smart building condition. This research also tries to investigate the mitigation by direction of arrival of the radio signal. Additionally, the wellness approach aims to generate the customised methodology that will support, suggest and assist the system designer in a smart building environment to evaluate and measure the on-site performance, so that the assessments are efficient, precise and accurate.

4.1 Introduction

A transmitting antenna radiates radio signal simultaneously in all directions; the signal usually chooses many different paths to arrive at the receiver. In each path, the signal interacts with communication environment ambiguously and reaches the receiver with some delay, and usually, a change in phase and frequency takes place. If the signals collected at the receiver are in phase, they produce the constructive interference. In case the signals are out of phase, they cause a loss in signal strength

H. Ghayvat and S.C. Mukhopadhyay, *Wellness Protocol for Smart Homes*,
Smart Sensors, Measurement and Instrumentation 24,
DOI 10.1007/978-3-319-52048-3_4

as they produce destructive interference. The direction of arrival (DOA) measurement produces the best approach to WSNs localization for the least multi-path loss. Ideally, wireless communication devices use omnidirectional low power wire whip as well as chip antenna that radiates radio signal consistently in all directions equally. The performance of practical antenna does not match with the ideal one. The direction of arrival has a compelling role in an indoor environment where the system finds various types of obstructions that degrade the performance. The orientation of antenna plays a major role in their performance. All investigations and characterizations of wireless radio communication have been made by the measurement of collected data. This data reveals the nature of EM wave and with the application of statistical assessment over this data a mathematical model and equation has been generated.

The aim is to perform experiments and contribute to building the mathematical model that include the real world effects and parameters. This study presents a realistic situation of a smart building monitoring system. The intelligent building monitoring can be defined as an application of automation and embedded technology with integral systems of accommodation facilities to enhance and progress everyday lifespan of an occupant. A wellness protocol based intelligent building monitoring system has been designed (Ghayvat et al. 2015a, b).

The rest of the research is organized as follows. Section 4.2 initiates with the Wellness-based smart building environment description. Section 4.3 presents the Methodology to measure interference and attenuation loss. Section 4.4 presents the practical issues based on interference and attenuation losses. Moreover, it describes the mitigation and possible suggestions to handle the issues. Lastly, Sect. 4.5 concludes the outcome.

4.2 Description of Smart Building System

The smart monitoring approach has first been implemented in an old house and later it is extended to a building apartment. The smart and intelligent monitoring system has been running continuously for the last 18 months without any major problem, in real world condition where it has occupants. In the wireless radio communication, different types of ZigBee RF modules are available. For the current research, the Digi XBee Series-2 RF has been selected. The design and deployment of the present sensing system inside the smart building are shown in Fig. 4.1. The layout depicts the locations of sensors in a smart building with occupants.

The heterogeneous sensing units include PIR, temperature and electronic and electrical appliances monitoring unit, force sensor and manual button.

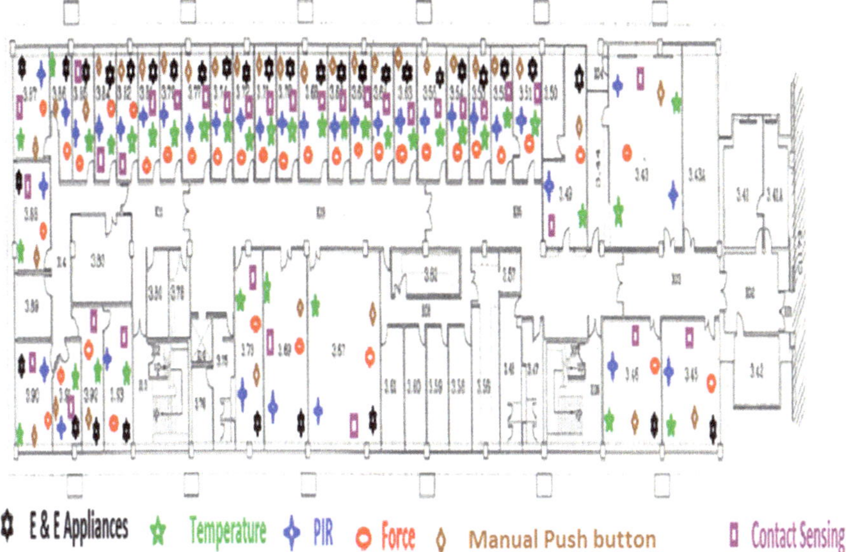

Fig. 4.1 Layout of big building heterogeneous sensing system

4.3 Methodology to Measure Interference and Attenuation Loss

The research approach proposed in this study has been implemented under an uncontrolled and realistic urban environment with simple and compact components. Since reliability is indispensable, the desired user data must be transmitted, received and analysed within allowable and considerable time duration in near real time with the best precision value and the least possible error.

In the WSN-based smart home, defining the reliability in system performance perspective is not easy because the radio communication system finds some variables that detrimentally influence the reliability metrics. The most common explanation for reliability is related to data parameters; these are data reliability, data transmission, reception, measured error, and delay. Data transmission and reception are the functions of the communication link between sensor unit. The quality of radio communication link boosts the chances of reliable data delivery. However, the design consideration of the sensor node also decides the performance because they are the source of data, and if data from the source is corrupted, the system cannot achieve a significant accuracy. Mostly, even with good RF link and robust sensing unit design consideration, the system is not accurate. The cause of this is improper routing selection and obstruction in the building environment. The home environment is full of obstructions, so it causes loss of data. Improper routing and topology selection increase this data loss because sometimes an end-device node does not find the nearby router node and data usually gets lost. The signal transmission takes place with the speed of light, so the delay is negligible. The gross delay at the sink will be significant enough to affect the near real time streaming (Ghayvat et al. 2015b; Liu et al. 2015).

This can be better understood by the packet reliability terms of Wellness protocol-based WSNs. Packet success rate (PSR), Packet delivery ratio (PDR), Packet error rate (PER), Packet loss rate (PLR), Signal to Noise Ratio, Received signal strength and received packet delay are basic parameters that define system reliability and performance. To evaluate these parameters in Wellness Protocol-based wireless sensor and networks, the smart building setup is designed and implemented in real-time applications. The results show that the deployment environment, distance, and localisation of sensing nodes are key factors that decide the reliability of wireless sensor and networks. For system performance evaluation and better understanding, these parameters have to be formulated with real world effects. The parameters are as follows.

4.3.1 Latency

In the wireless system, the real-time performance is highly affected by traffic status, collision and congestion. If the sensed information is sent from the sensing nodes at a predefined sampling rate, it is expected that the receiver receives the information at the same rate. These packets are routed through various communication paths, so they take different time, and some of them may be lost in the endeavor to the destination. There is the provision of time out, i.e. the deadline, for these packets' arrival. Hence, one of the great pointers of wireless network performance is the packet delay between consecutive packets perfectly received by the Coordinator. To assure the temporal rationality of time-signature, the measurements need to be executed on the same computer system and with the common clock reference.

The wellness protocol allows sensing nodes to transmit data in a peer-to-peer, point-to-point or point-to-multipoint (star) network architecture. The time it takes to transmit a data packet is a sum of the Time on the Air, an acknowledgment (Time for CSMA-CA and Retries) and propagation time (Ghayvat et al. 2016).

Annotation:
B bit/bytes
Sd_b Starting Delimiter
Add_b Address bytes
Pl_b Payload bytes
C_b Cyclic redundancy checks bytes
Pc_b Packet control bytes
Adr_{bs} Air data rate in bit per sec

$$\text{Total received delay time } (T_{dt}) = \text{Time on the air}(T_a) + \text{Time on-air Ack}(T_{ack}) + \text{Time upload } (T_{sci}) + \text{Propagation time } (T_p) \tag{4.1}$$

$$T_a = \frac{8(B)\{[Sd_b] + n_2[Add_b] + N[Pl_b] + [C_b] + n_1[Pc_b]\}}{Adr_{bs}} \tag{4.2}$$

$$T_{ack} = \frac{8(B)\{[Sd_b] + n_2[Add_b] + N[Pl_b] + [C_b] + n[Pc_b]\}}{Adr_{bs}} \tag{4.3}$$

$$T_{sci} = \frac{8[B].\{[Pl_b]\}}{Adr_{bs}} \tag{4.4}$$

The propagation delay T_p is the time that takes a signal to propagate through the communications media from a sensing node to the next sensing node. It can be computed using the following equation.

$$T_p = D/S \tag{4.5}$$

Here, D is the distance from the node to the next node and S is the propagation speed of the media. The latency calculated from Eq. 4.1 includes some extra overheads caused by the computational and communication systems and controllers. To reduce the jitter, it is recommended to limit the processing load on the computer. The jitter is a complex issue usually related to the wireless networks and infrastructure. This delay is closely linked to inter-packet delay. Inter-packet delay is the degree of the inconsistency over the wireless communication time of the latency across a wireless network. One of the best possible solutions is to introduce the jitter buffer to queue up the received message. In the wireless network-based smart building, instead of introducing jitter buffer between network and converter, it can be done by utilizing the RAM capacity efficiently. This reduction in load results in terms of availability of RAM for packet transmission and reception without I/O operations blocking.

4.3.2 Data-Packet Delivery Parameters

The packet delivered from the source node is expected to be received at sink node without any distortion, error, and too much delay that degrades the accuracy. This is usually evaluated by the number of messages sent from the source and received at the destination gateway. The experiments are implemented by regularly transmitting and receiving packets from the transmitter Tx (sensor node) to the receiver (coordinator) Rx.

PDR, PLR, PER, and PSR are directly linked to the system performance and packet reliability. They represent the packet accuracy at various levels.

- PDR: Packet delivery ratio is the number of data packets received at the gateway-coordinator to the number of data packets sent from the transmitter node. It is represented in percentage.

$$\mathrm{PDR}\ (\%) = \frac{\mathrm{N_r}}{\mathrm{N_s}} * 100 \tag{4.6}$$

N_r a total number of data packets received by the coordinator from an end-device.
N_s a total number of data packets sent by the end-device.

The more the PDR value, the better the accuracy.

- PSR: Packet success rate is the number of data packets successfully received without any error to the number of data packets received including errors at the gateway-coordinator. It is represented in percentage.

$$\mathrm{PSR}\ (\%) = \frac{\mathrm{N_{we}}}{\mathrm{N_r}} * 100 \tag{4.7}$$

N_r a total number of data packets received by the coordinator from an end-device.
N_{we} a total number of data packets received by the coordinator without error from an end-device.

The more the PSR value, the better the accuracy.

- PER: Packet error rate is the ratio of the number of data packets received with error, and the number of data packets received at the gateway coordinator. It is represented in percentage.

$$\mathrm{PER}(\%) = \frac{\mathrm{N_r} - \mathrm{N_{we}}}{\mathrm{N_r}} * 100 \tag{4.8}$$

N_r a total number of data packets received by the coordinator from an end-device.
N_{we} a total number of data packets received by the coordinator without error from an end-device.

The lesser the PER value, the better the accuracy.

- PLR: Packet loss rate is the ratio of the number of data packets received from the coordinator, and the number of data packets sent from the transmitter. It is represented in percentage.

$$\mathrm{PLR}(\%) = \frac{\mathrm{N_s} - \mathrm{N_r}}{\mathrm{N_s}} * 100 \tag{4.9}$$

N_r a total number of data packets received by the coordinator from an end-device.

N_s a total number of data packets sent by the end-device.

The lesser the PLR value, the better the accuracy.

4.3.3 Link Quality Metrics

Electromagnetic interference is the disruption that upsets desired sensing node signal processing through electromagnetic radiation emitted from an external source. ZigBee-based sensor nodes operate in the 2.4 GHz ISM spectrum band. The wellness system uses XBee S-2, and it is based on ISM band. The causes of disturbance in this frequency spectrum are:

- Bluetooth (IEEE 802.15.1)
- Microwave oven
- Wireless USB (IEEE 802.15.3)
- WI-Fi (IEEE 802.11)
- Some other sources like RF motion detectors and cordless phones.

The effect of this disturbance can be examined by received signal strength and Signal to Noise Ratio (SNR).

4.3.3.1 The Received Signal Strength Indicator (RSSI)

The Received Signal Strength Indicator (RSSI) is the level of the received signal at Gateway coordinator in –dBm from a sensor node. The strength of the detected signal reduces with increase in spacing between the transmitter and the RSSI also depends on path loss effect. Additionally, the quality of the signal is affected by other wireless devices operating in ISM band. If the transmitting node has a transmission power of P_t Watts and this antenna power in the logarithmic domain is expressed in dBm (in mW). With an isotropic antenna gain of G_t dBi, the total effective isotropic radiated power (EIRP) is $P_t * G_t$. Therefore, the received signal at the receiver is given by (Ghayvat et al. 2016)

$$P_r = P_t + G_t + G_r - PL \tag{4.10}$$

G_r is the gain of the receiver antenna, and PL is the combined path loss parameters that add attenuation due to path losses.

In the building environment, the path loss is represented by

$$PL = L_{fs} + L_{sf} + L_{ff}(t) \tag{4.11}$$

L_{fs} is the term for free space Line-of-Sight (LOS) path loss.

L_{sf} loss is due to slow fading caused by large static obstructions such as a wall. L_{ff} loss is the small-scale fast fading loss that happens due to destructive interference from multipath effects.

Destructive interference is the primary and most significant loss in the building environment. This loss varies with time. Most of the time in the building ambiance, the electromagnetic signal does not find the LOS between transmitter and receiver, so it travels by multipath or in non-line-of-sight direction. Minimum acceptable SNR value $SNR_{P(T,R)}$ should be greater than the threshold value, η to recover the transmitted signal. The threshold value is a signal from which signal can be extracted. The value of η is defined by multiple experiments and measurement in that particular environment.

$$SNR_{P(T,R)} = SNR_{Pref} - 10\beta \log \{\frac{(d(T,R))}{d_{ref}}\} + X_{\sigma} + \alpha(T,\ R) \qquad (4.12)$$

Propagation loss estimation in a building environment shows that the average strength of the received signal at any point in the network decays as a power law of distance $d(T,R)$ between a transmitter and a receiver. SNR_{Pref} is the power received at a close reference point in the small reference distance from the transmitting node and β is the path loss exponent. X_{σ} is zero-mean gaussian distributed random variable with standard deviation σ in dBm. It is used for long term variability consideration, which is used to approximate the fading phenomenon in building environment. α(T, R) = attenuation in dB per unit distance which varies according to material and obstruction type. The term α(T, R) and β need to be determined in each wireless network individually.

An isotropic antenna is known for radiating the power uniformly in all directions (in a sphere), which is only possible in theory and difficult to achieve in actual practice. The Direction of Arrival (DOA) of the signal at the receiver becomes significant in the building environment where wireless sensing system finds obstructions and interference sources. There is a direct relationship between the direction of a signal and the associated received steering vector. In actual practice, defining and evaluating the DOA is complicated because there are unknown numbers of RF signals striking on the receiving antenna concurrently, each from unidentified directions and with unknown amplitudes. Additionally, the received RF signals are always degraded by the disturbance caused by noise, attenuation, and interference. The expression of the received signal can be written as (Ghayvat et al. 2015c),

$$P_{(T,R)} = P_R + \sum_{j=1}^{N} P_{ij} \cos(\theta_j + \alpha_i) + \sum_{j=1}^{M} P_{pj} \cos(\theta_j + \beta_p) \qquad (4.13)$$

P_R Peak or reference received power value at the coordinator end.
P_{ij} Power affected by interference sources, $_j$ is the number of interference sources. It is 1 to N.
P_{pj} Power affected by multi path fading, $_j$ is the number of multi path loss sources or multi path fading level. It is 1 to P.
θ_j Angle of arrival with respect to interference sources.
α_i Phase change caused by interference sources.
β_p Phase change caused by multi path fading.

4.3.3.2 Signal to Noise Ratio (SNR)

The other devices, which are operated in ISM band, cause disturbance to the desired RF signal. These undesired disturbance sources produce RF power that interrupts the required signal for smart home operation. This is ultimately referred to as noise.

$$\mathrm{SINR} = \frac{P_{Signal}}{P_{noise} + P_{interference}} \tag{4.14}$$

$$\mathrm{SNR}\,(\mathrm{dBm}) = P_{Signal} - P_{noise} - P_{interference} \tag{4.15}$$

4.4 Experimental Observations, Analysis, and Mitigation

The experiments and analysis are done in the following steps.

4.4.1 Fundamental Tests

Packet delivery parameters (PER, PLR, PSR, and PDR) and link quality parameters as a function of spacing. These are the basic tests conducted to understand the performance of the wireless sensors and networks in the smart building environment, with minimum interference. All the experiments and setup have been developed to generate the plots for each metric as a function of distance. The sensing nodes are deployed in the smart building, with different spacing between the nodes. For analysis, over 5000 samples are collected to define the uncertainties. The transmitter power is set at 2 mW (+3 dBm) boost mode, and the receiver sensitivity is −96 dBm.

The effect of the hopping distance on the arrangement of sensor nodes is shown in Fig. 4.2. Figure 4.3 shows the delays in the communication as a function of hopping distance. Hop1 means the packet is sent without any router device while hop2 represents the packet is sent to a router between the end device and the

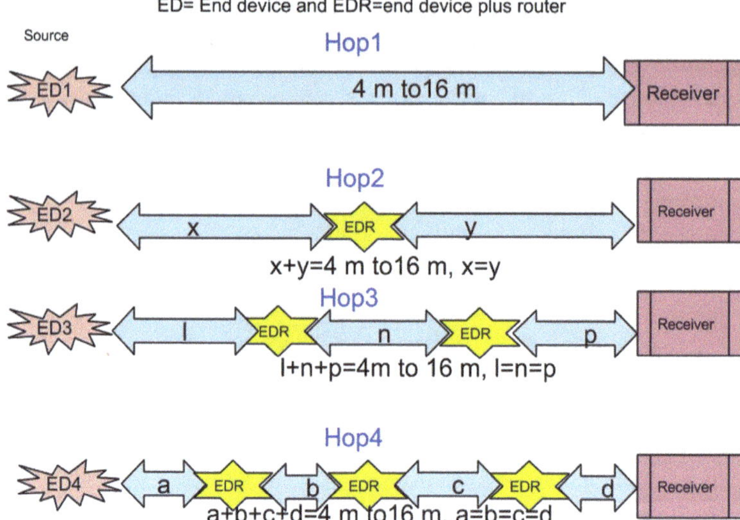

Fig. 4.2 Arrangement of sensor nodes for experimental investigation

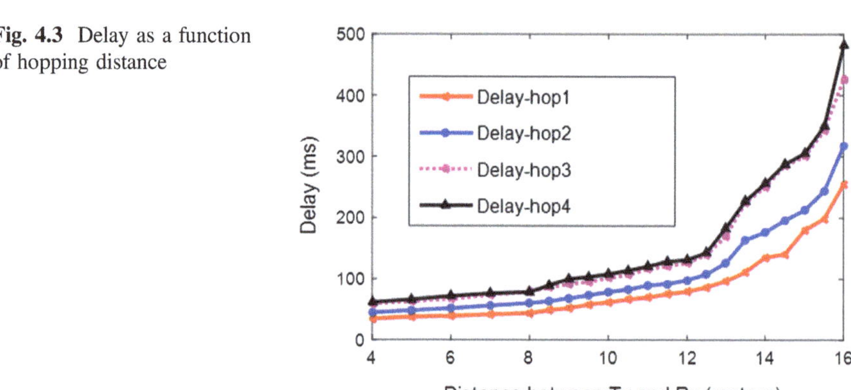

Fig. 4.3 Delay as a function of hopping distance

coordinator. Similarly, hop3 and hop4 show the packet is sent via 2 and 3 router devices between the end device and the coordinator, respectively. In hop1, the spacing between the Transmitter-end-device (Tx) and the Receiver-coordinator (Rx) is varied. Whereas in hop2, hop3 and hop4 the end-device and the coordinator are kept at a defined distance and the routers spacing with respect to the end-device and the coordinator is varied. Hop2, hop3, and hop4 are used as router-end-devices. The delay has increased with the increase in the spacing between source and destination. Additionally, the number of delays increases with using a large number of hops. Every time the end-device (ED) delivers the packet to the end-device-router (EDR) in the case of multi-hop communication and this EDR keeps its own packet delivery at the highest priority. If the EDR generates its own packet P1 and at the same moment, it receives relay packet P2 from another neighbor node, it will deliver P1 followed by P2, which results in the latency. The estimated uncertainties in delay are 27 ms based on the standard deviations of multiple investigations. Although the delay value that the server has recorded is in milliseconds, it has the potential to change the real-time application to near real-time application. The whole setup is the WSNs based realistic, smart building environment where some of these sensors are event based, and the rest have fixed sampling rate. The length of the packet varies from 20 to 24 bytes.

Figure 4.4 shows the PDR values of all four different setups. In all cases, PDR decreases with increase in spacing. In three cases, PDR drops in PDRHop3, 2 and 1, except PDRHop4 where the system has used 3 EDRs. In PDRHop1, the drop is quite significant where it starts decreasing after 8 m followed by PDRHop2 after 12 m and PDRHop3 after 14 m. The estimated uncertainties in PDR are 1% based on the standard deviations of multiple investigations.

Figure 4.5 shows the PER values of all four different setups, in all the cases PER increases with increase in spacing. In all four cases, PER raises significantly after the spacing 8 m. In PERHop1, the drop is quite major where it starts increasing,

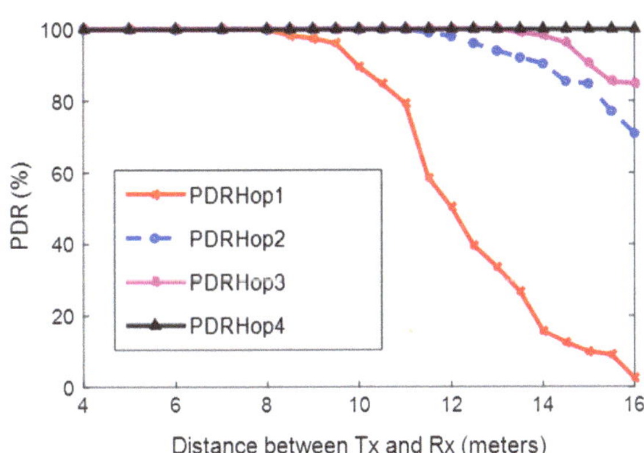

Fig. 4.4 PDR as a function of hopping distance

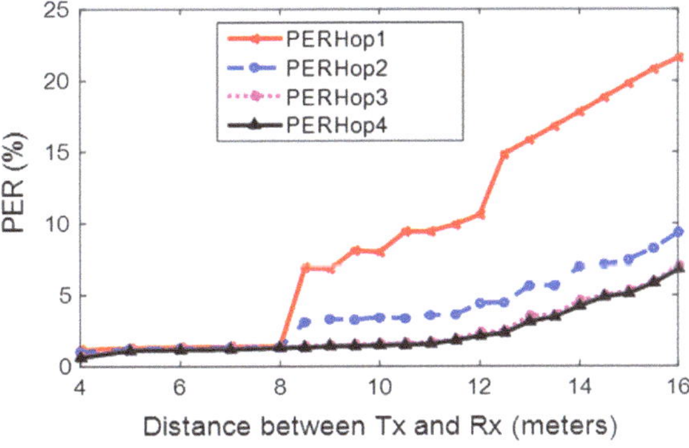

Fig. 4.5 PER as a function of hopping distance

followed by PERHop2. While PERHop4 has been the least affected followed by PERHop3. The estimated uncertainties in PER are 1.37% based on the standard deviations of multiple investigations.

By increasing the spacing between Tx and Rx the packet delivery performance degraded, so a better solution is to deploy the EDR to improve it.

4.4.2 Analytical and Parametric Tests

Experiments have been conducted to assess the effect of various parameters on packet delivery parameters (PER, PLR, PSR, and PDR) and link quality parameters. The issues investigated in this research work are as follows:

1. Multipath fading and spacing between Tx and Rx (S_{Tx-Rx});
2. Interference Source (IS) and spacing between IS and Rx (S_{IS-Rx}), and IS and Tx (S_{IS-Tx});
3. The direction of arrival (DOA) in relation to IS to the line-of-sight and multipath fading, additionally, the angle between Rx and IS (θ) is considered.
4. Communication channels.

The aim of these tests is to investigate the effects of different issues on the performance of WSN for Smart building. Table 4.1 presents the issues and the range of tests. In IS, the research has considered the most common sources of disturbance in ISM band. In the endeavor to find the least affected channel with Wi-Fi, a study has selected 15, 25 and 26 (Ghayvat et al. 2016). These three channels get least affected by Wi-Fi existence. The distance between the ED and the coordinator is kept from 2 to 16 m, which is sufficient to consider one floor of the building apartment. To observe the effect of IS on received signal, the spacing between IS and Rx varied from 0.5 to 5 m. This effect has been recorded for line-of-sight and multipath fading environment.

Table 4.1 The issues and their range in the experiments

Number	Issues	Range
1	Interference sources (IS)	Wi-Fi, Bluetooth, microwave, cordless phone
2	Communication channel	15, 25 and 26
3	S_{Tx-Rx}	2–16 m
4	S_{Is-Rx}	0.5–5 m
5	Deployment angle of IS with respect to Tx–Rx line of sight	0°–360°
6	Multipath fading materials by attenuation source (AS)	Glass (GLS), plywood (PYWD), metal (MTL), concrete (CNCRT), brick (BRCK), coated glass (CTDG)
7	Separation S_{Tx-AS}	0.5–6 m

The antenna chip/wire/whip are isotropic antennas so the direction of arrival of a signal and their location with respect to receiver and IS should not affect the performance ideally. In actual practice, the analysis shows that the location and the DOA have considerable consequences. In the experimental setup, the IS kept at different location and angle from 0° to 360°. The living environment contains heterogeneous sensors, and the deployment of these sensing units is decisive to get optimum received signal value. The building is made of different materials and contains various household objects (steel chair, metal table, etc.) which cause the attenuation loss to low-power wireless communication. This research study includes all possible sources of attenuation that degrade the signal characteristics.

4.4.3 Signal Attenuation Inside Smart Building

The communication performances inside smart buildings are examined to mitigate the losses and generate trade-off in the deployment environment. WSNs offer remarkable prospects for effective utilization of power and enrichment in occupant's well-being in the building environment by improving the communication link quality to make sensing data available all the time. A major issue in the implementation is the ambiguity between designers concerning the consistency of wireless communication links via building material and household objects. Tests have been done to investigate the RSSI and packet delivery parameter values as a function of transmitter-attenuation source separation S_{Tx-AS} distance and obstruction type. These experiments are implemented through packet reliability, the switch from consistent to inconsistent wireless communications at different distances and materials. The packet delivery rate is measured in the building; the RSSI is interrelated with the packet success rate in an uncontrolled noise environment. These tests were carried out last year, with 5000 plus samples analyzed in each test. Table 4.2 shows the independent constraints variations in the experiments (Ghayvat et al. 2016).

For this particular test, the fixed sampling rate of 5 ms has been selected, and packet size of 22 bytes has been kept. The distance between S_{Tx-Rx} has been maintained at 8 m, and the separation between obstructions to transmitter varied

Table 4.2 Obstruction-material details of building environment

Material	Thickness	Remark
Plywood as well as wood	2.3 cm	Doors as well as divider
Metal mostly steel/iron	1.2 cm	Doors, tables, bed, and chairs, interior office panel
Concrete	27.5 cm	Wall
Brick	29.3 cm	Wall
Glass and coated glass	1.2 cm* and 1.7 cm	Window* and doors, divider

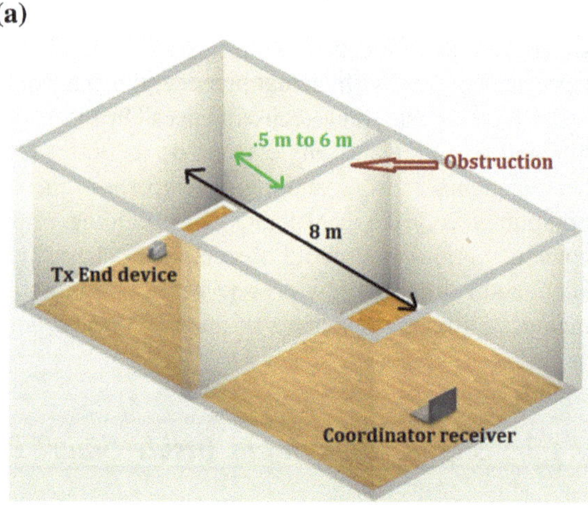

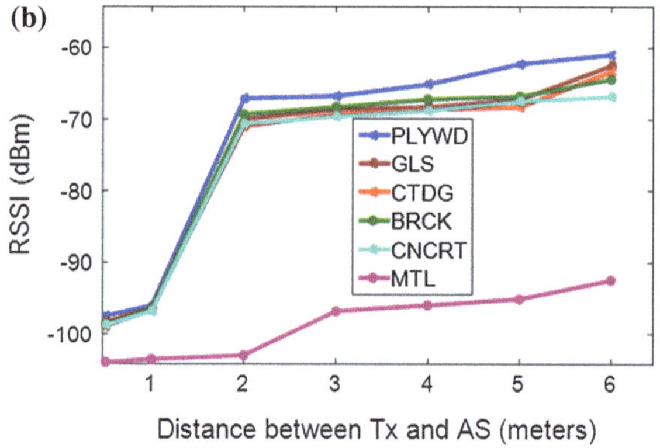

Fig. 4.6 **a** Layout of attenuation loss test, **b** RSSI as a function of different building materials, where S_{Tx-AS} is varied from 0.5 to 6 m

from 0.5 to 6 m. Figure 4.6a presents the experiment at the layout of the smart building. The tests have been conducted at various obstruction conditions, which are present in the smart building, to examine the adverse effect of coated glass, concrete, metal, plywood, brick and glass on the wireless signal multipath fading. The effective area of obstructions has been varied.

The repeated experiments were all carried out in the open field as well as inside a controlled environment. In an environment like a building, there are many sources which may influence the system. In order to generate accurate results from analysis and observations, all the possible considerations of dependent and independent variables have been made. Figure 4.6b shows the average RSSI values as a function of various obstructions while this separation between Tx and AS varied from 0.5 to 6 m. The measured standard deviation (σ) is ± 3 dBm. If the separation between Tx and AS is less, the RSSI value will be poor. Metal has caused major degradation in signal propagation. For instance, the variation from 0.5 m to 3 m between Tx and metal AS, decreases the RSSI from -104 to -96.81 dBm, which is not good enough to recover packet from it. Plywood and wood are the least affecting AS followed by glass, brick and concrete, and finally coated glass. Up to S_{Tx-AS} = 1.5 m, the RSSI for AS glass-coated glass, brick, and concrete were less than the threshold sensitivity level of the receiver. With the spacing $S_{Tx-AS} \geq 2$ m, the RSSI improved significantly in the glass, coated glass, brick, and concrete.

Figure 4.7a, b show the PDR data. The graphs represent that nearly all AS PDR values have secured over 93% except metal AS for the spacing $S_{Tx-AS} \geq 2$ m. With the increase in S_{Tx-AS} value, the PDR values in all AS have been improved considerably. The PDRMETAL has recorded over 90% for $S_{Tx-AS} > 3$ m. There is a slight difference in the PDR metrics of glass, plywood, concrete, and coated glass. The major improvement in the PDR with an increase in separation is in plywood, followed by glass, brick, and finally, concrete and coated glass. The reliable range of communication in the glass, plywood, concrete, and coated glass is $S_{Tx-AS} \geq 2$ m, while metal is an exception. The measured standard deviation (σ) for PDR is 2%.

Figure 4.8a, b follow the same trend which is shown in PDR. Except for the PSRMETAL, the rest have touched over 96% for spacing $S_{Tx-AS} \geq 2$ m. The measured standard deviation (σ) for PSR is 1.5%.

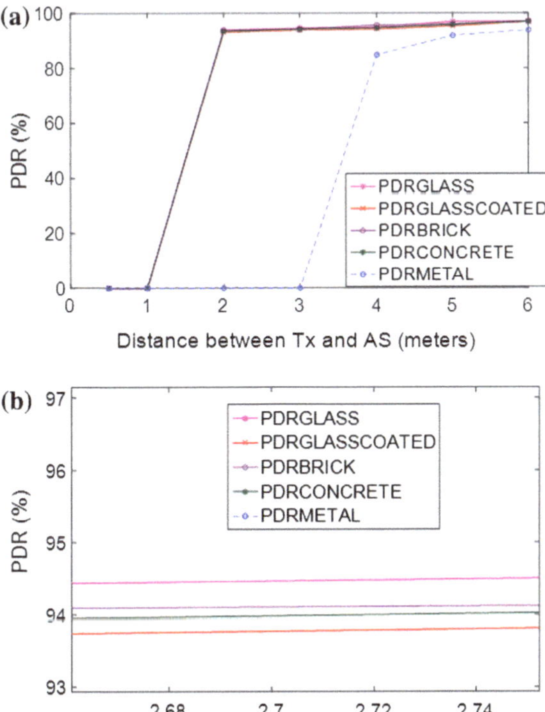

Fig. 4.7 **a** PDR as a function of different building materials, where S_{Tx-AS} is varied from 0.5 to 6 m, **b** close-up view PDR as a function of different building materials, where S_{Tx-AS} is varied from 0.5 to 6 m

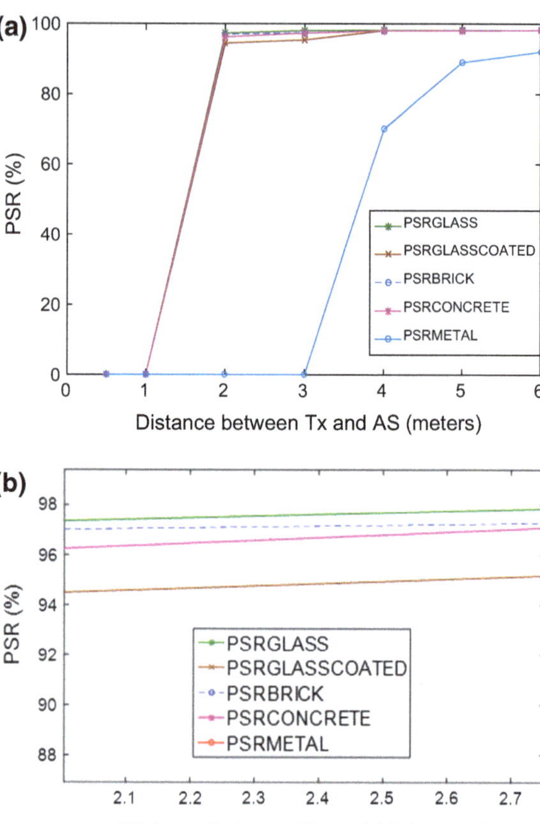

Fig. 4.8 **a** PSR as a function of different building materials, where S_{Tx-AS} is varied from 0.5 to 6 m, **b** close-up view PSR as a function of various building materials, where S_{Tx-AS} is varied from 0.5 to 6 m

Despite the large uncertainty in the recorded data, analysis has shown some clear trends. Wood, plywood, and glass cause little disturbance to the wireless signal propagation, and this is justified by the low dielectric constant values of these materials. Wellness approach has been developed to evaluate the possible ways to avoid unreliable communication as well as generate trade off to get optimum packet reliability parameters. Table 4.3 shows the optimum values for various attenuation sources.

Table 4.3 Optimum value recorded at $S_{Tx-AS} = 6$ m

Parameter at a distance 6 m	Plywood	Glass	Glass ctd	Brick	Concrete	Metal
RSSI	−61	−62.49	−63.32	−64.4	−66.88	−92.47
PDR (%)	97.55	97.11	97	97.02	97	94
PSR (%)	98.15	98.15	98.15	98.15	98.15	92

4.4.4 Direction of Arrival (DOA)

These tests demonstrate how the direction of arrival (DOA) can help to get the best signal performance as a function of Interference Source (IS). In the presence of IS, the system does not get optimum signal performance, which is highly essential for packet reliability. When IS is placed, it influences and causes noise to the desired wireless signal. This noise is reflected in terms of packet loss as well as an error. The aim of the research is to investigate the possible approach and methodology to mitigate it.

For this test, one wireless transmitter and one receiver coordinator have been arranged. Various IS sources are deployed, and their spacing from receiver have been varied. Moreover, the angle between the Tx-IS and Rx-IS have been varied to investigate the best possible deployment angle in the particular environment. Figure 4.9 presents the schematic setup of DOA measurement. The tests have been conducted in two different setups. The first setup is for the line-of-sight and the second setup is for multipath fading. These two experiments are performed for three different spacing $S_{Rx\text{-}IS}$ = 1 m, 2 m and 3 m. Additionally, in every experiment either Tx or Rx angle is varied as a function of IS location. In line-of-sight, the desired wireless signal is majorly affected by IS interference while in multipath the wireless signal is affected by AS and IS. The RSSI values are received for three different locations of IS as a function of Rx, Pri represents the RSSI value when the Rx angle has been varied with respect to IS while Ptr shows the RSSI value when the Tx angle has been changed with respect to IS location.

Figure 4.10 presents the RSSI values as a function of the angle between Rx and IS in line-of-sight setup, where the spacing between Tx and Rx is fixed at 3 m. The RSSI values have been recorded for three spacings Pri1 (when Rx is 1 m from IS), Pri2 (when Rx is 2 m from IS) and Pri3 (when Rx is 3 m from IS). In the particular test the peaks in RSSI values have been achieved at 60°, 150°, 240° and 330°. For the same angle, PDR and PSR values have touched the peak, as shown in Figs. 4.11 and 4.12 respectively. The more the spacing between Rx and IS, the lesser the interference effect on desired wireless signal, and the better the signal

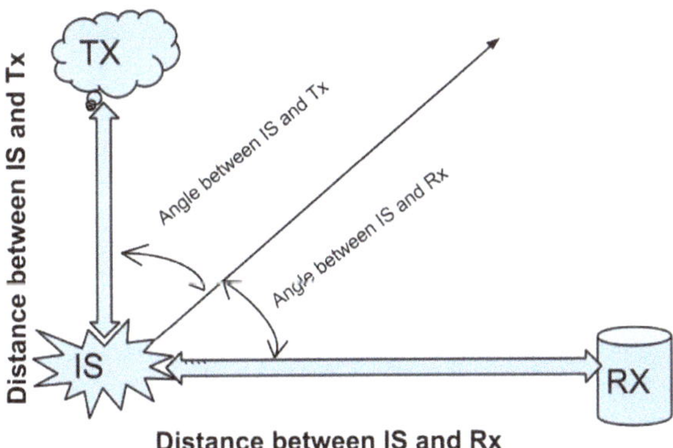

Fig. 4.9 Schematic setup for IS location for DOA

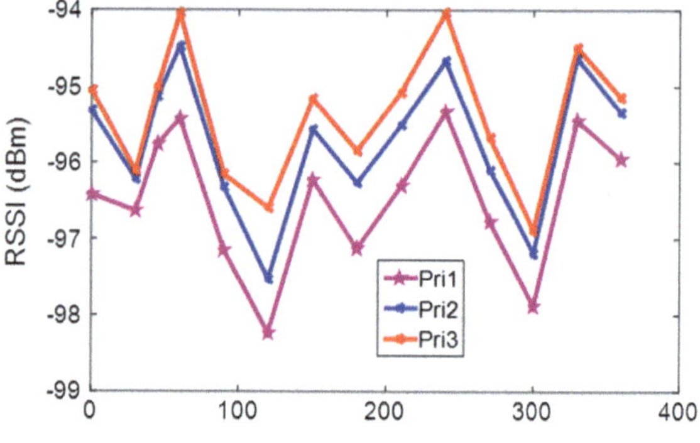

Fig. 4.10 RSSI as a function of the angle between Rx and IS in the *line of sight*, S_{Tx-Rx} = 3 m

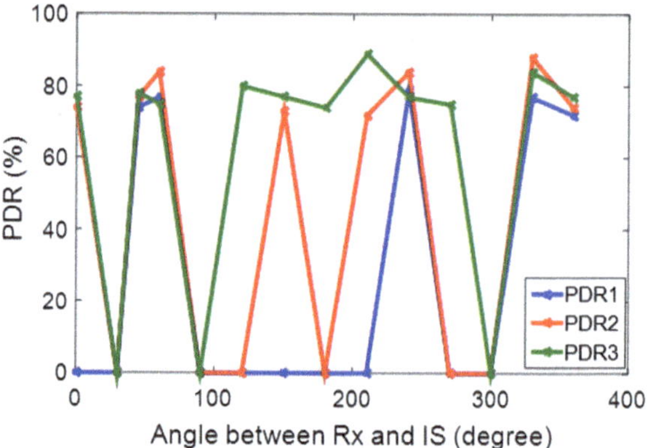

Fig. 4.11 PDR as a function of the angle between Rx and IS in the *line of sight*, S_{Tx-Rx} = 3 m

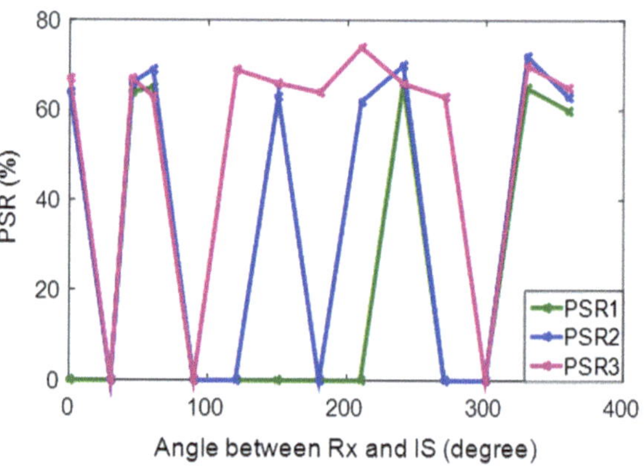

Fig. 4.12 PSR as a function of the angle between Rx and IS in the *line of sight*, S_{Tx-Rx} = 3 m

performance and packet reliability. Pri3, PSR3, and PDR3 have shown very good results while Pri1, PSR1 and PSR1 are affected considerably. The highest RSSI value has been recorded at 60° which is −94.01 dBm and packet metrics PDR is 87.46% and PSR is 76.28%. The measured standard deviation (σ) for RSSI is ±1 dBm, PDR is 2%, and PSR is 1.5%.

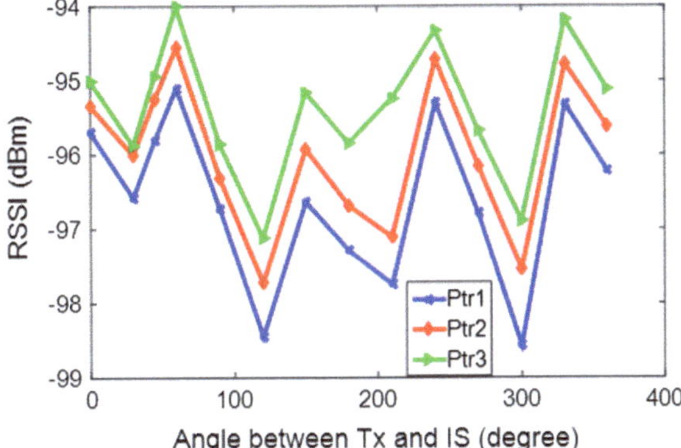

Fig. 4.13 RSSI as a function of the angle between Tx and IS in the *line of sight*, $S_{Tx-Rx} = 3$ m

Figure 4.13 presents the RSSI values as a function of the angle between Tx and IS in line-of-sight setup. In the particular test, the best possible RSSI values have been achieved at 60°, 150°, 240° and 330°. For the same angle, PDR and PSR values have attained high values, as shown in Figs. 4.14 and 4.15 respectively. Similar trends of IS and Rx spacing have been recorded in these graphs as well.

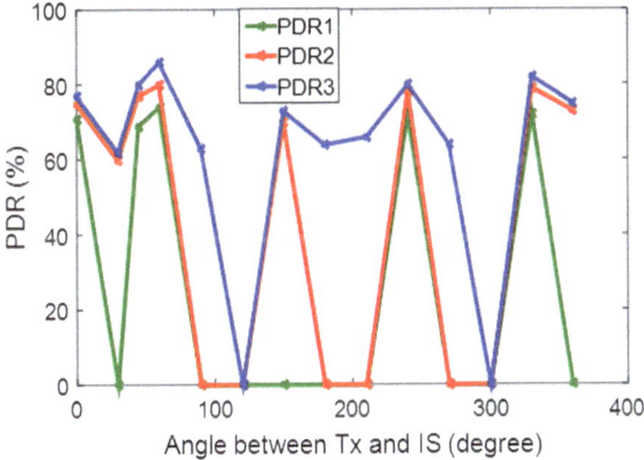

Fig. 4.14 PDR as a function of the angle between Tx and IS in the *line of sight*, $S_{Tx-Rx} = 3$ m

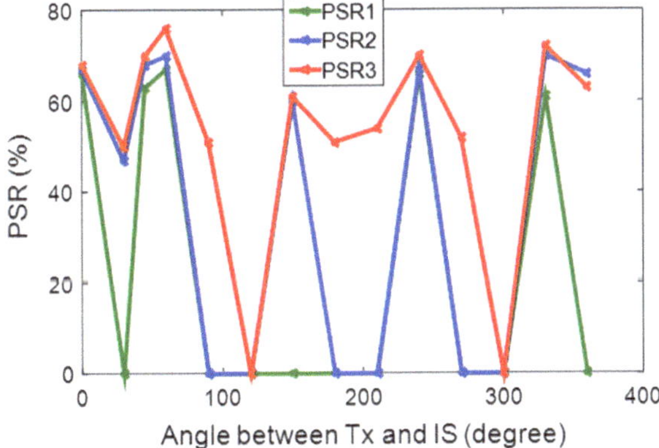

Fig. 4.15 PSR as a function of the angle between Tx and IS in the *line of sight*, $S_{Tx-Rx} = 3$ m

The highest RSSI value has been recorded at 60° which is −94.04 dBm and packet metrics PDR is 86.35% and PSR is 75.29%. The measured standard deviation (σ) for RSSI is ±1 dBm, PDR is 2% and PSR is 1.5%.

Figure 4.16 presents the RSSI values as a function of the angle between Rx and IS in multipath setup, where the spacing between Tx and Rx is fixed at 5 m, and there are some obstacles in between them. The RSSI values have been recorded for three spacing Pri1 (when Rx is 1 m from IS), Pri2 (when Rx is 2 m from IS) and Pri3 (when Rx is 3 m from IS). In the particular test, the peak in RSSI values has been achieved at 45°, 120°, 210° and 330°. For the same angle, PDR and PSR values have attained high values as shown in Figs. 4.17 and 4.18 respectively. The highest RSSI value has been recorded at 210° which is −94.61 dBm and packet metrics PDR is 80.21% and PSR is 71.18%. The measured standard deviation (σ) for RSSI is ±1 dBm, PDR is 2%, and PSR is 1.5%.

Similar trends were found in Fig. 4.19 that presents the RSSI values as a function of the angle between Tx and IS in multipath setup. In the particular test, the peak in RSSI values has been achieved at 45°, 120°, 270° and 330°. For the same angle, PDR and PSR values have touched high values as shown in Figs. 4.20 and

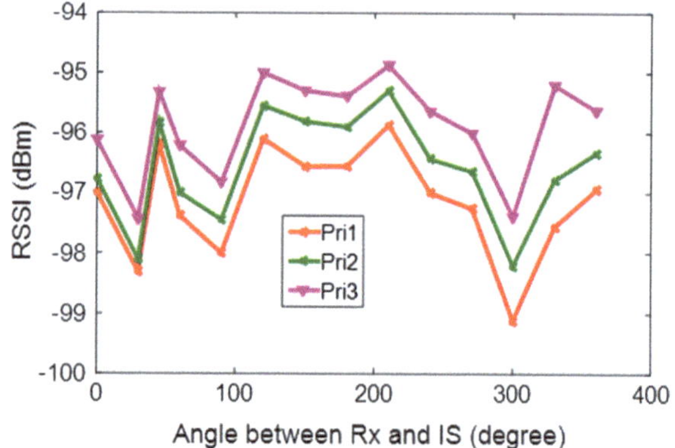

Fig. 4.16 RSSI as a function of angle between Rx and IS in multipath fading, $S_{Tx-Rx} = 5$ m

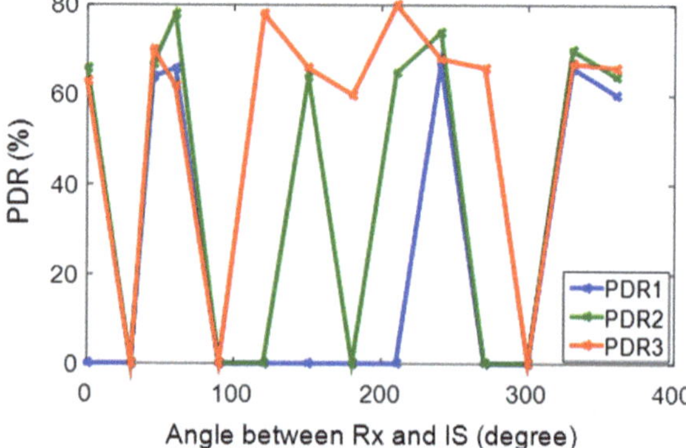

Fig. 4.17 PDR as a function of angle between Rx and IS in multipath fading, $S_{Tx-Rx} = 5$ m

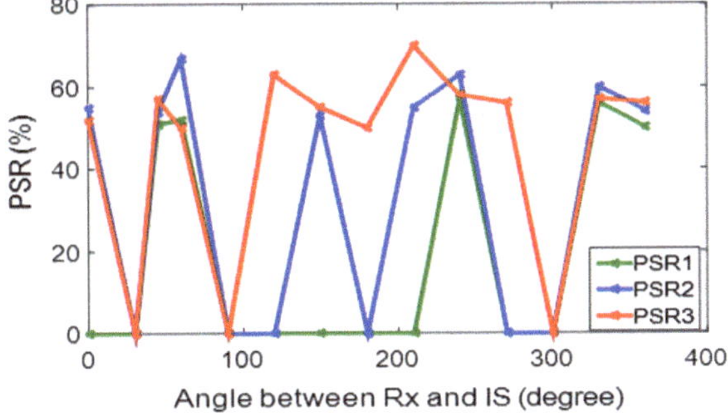

Fig. 4.18 PSR as a function of angle between Rx and IS in multipath fading, $S_{Tx-Rx} = 5$ m

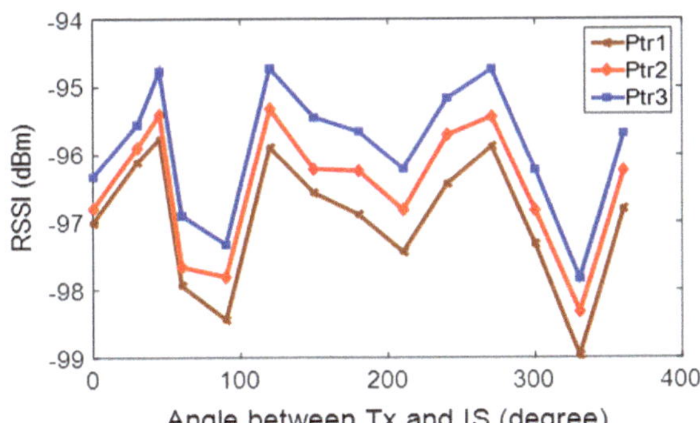

Fig. 4.19 RSSI as a function of angle between Tx and IS in multipath fading, $S_{Tx-Rx} = 5$ m

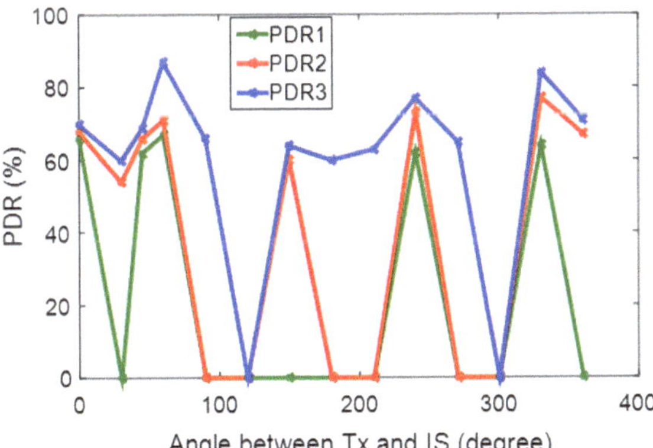

Fig. 4.20 PDR as a function of angle between Tx and IS in multipath fading, $S_{Tx-Rx} = 5$ m

4.21 respectively. Through these tests of DOA, it is clearly understood that direction of arrival of signal and spacing between the receiver and IS play significant roles on the interference effects of the radio communication signals.

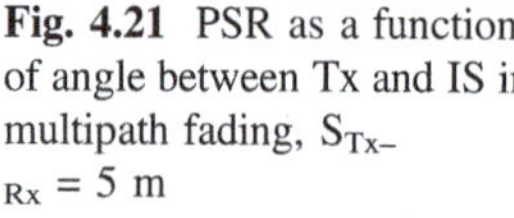

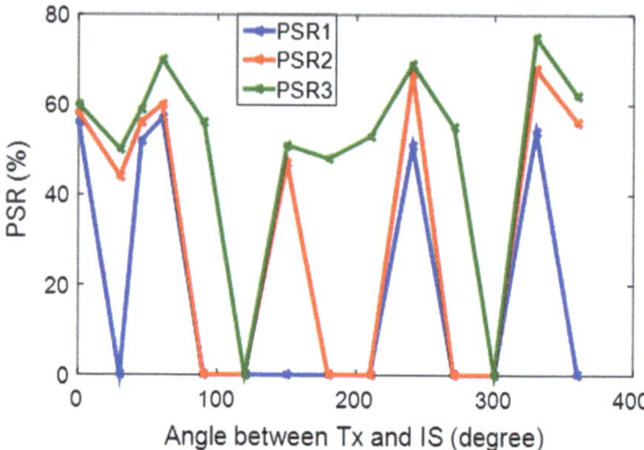

Fig. 4.21 PSR as a function of angle between Tx and IS in multipath fading, $S_{Tx-Rx} = 5$ m

4.4.5 *Mitigation of Interference and Suggestions*

Bluetooth, IEEE 802.11 b/g, IEEE 802.15.4, 2.4 GHz frequency hopping spread spectrum portable phones and a number of proprietary wireless technologies operate in the ISM band. The coexistence of different technologies with Wellness protocol has been captured using the exceptional and economical Wi-Spy tool and displayed using the Chanalyser package. The channel numbers along the bottom position are Wellness channels. Current research used MetaGeek Spectrum Analysis Wi-Spy DBx & Chanalyser 5 device as a frequency spectrum analyzer and configured it for a particular application requirement (Metageek 2016). Research analysis is using density and waterfall graphs to visualize the interference and loss caused by other RF device in ISM band. These graphs represent the RSSI values at different Wellness channels for the smart home monitoring system.

The Density View graph shows current activities within their loudness and frequency of transmission, the spectrum allowing to recognize devices. With 'Color by Utilisation' enabled, the height of the graph indicates loudness of the devices and the intensity of the color presents the frequency of occurrence. The more concentrated the color, the more often the frequency is in use. This is called utilization, which is analogous to duty cycle and airtime usage. For example, if the frequency has 40% utilization, it is only free for use by other transmitters for 60% of the time duration. A blue spike or profile shows a short signal, like a clap. A red spike or shape shows a long, unbroken-continuous signal, as an air horn. Colors and their significance are as follows: Blue—less than 10% utilization, Green—20% utilization, Yellow—40% utilization, Red—over 50% utilization.

The Waterfall View graphs amplitude over time for all frequencies in the selected band, much like a seismometer graph for earthquakes. This view is useful for watching the spectrum over time. Unlike the Density View, which uses Color by Utilisation, the intensity of the color in the Waterfall View shows amplitude. Blue indicates low-amplitude signals, while red indicates high-amplitude signals.

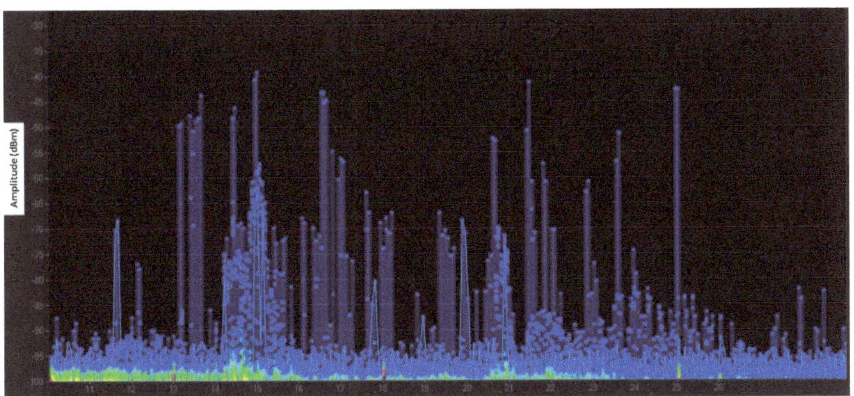

Fig. 4.22 The XBee smart building system is operating at frequency 2.430 MHz

Figure 4.22 shows the ZigBee-based system operating at channel 15 without any potential source of interference. Figures 4.23, 4.24 and 4.25 show the interference caused by other wireless technology in smart home monitoring Wellness channel.

The ZigBee-based WSNs are functioning at channel 2.430 GHz. The sensor nodes are deployed in Mesh topology into a smart building with spacing between nodes up to 6 meters, and the Wi-Fi, Bluetooth and microwave sources are placed at 2-meter distance from the receiver (coordinator). Figure 4.22 shows the density and waterfall views of XBee smart building system that is operating at frequency 2.430 MHz under minimum interference condition. Figure 4.23 shows the Wi-Fi operation that affected the XBee operation badly. Figure 4.24 shows the Bluetooth functioning over the same frequency of 2.430 MHz, which corrupted the XBee RF link quality. Figure 4.25 represents microwave oven signals dissipate across the whole Wellness spectrum and spoil the signal performance and packet delivery notably. This interference effect can be better understood by packet reliability parameters, shown in Figs. 4.26 and 4.27. The packet reliability metrics are most affected in microwave oven followed by Wi-Fi, and then by Bluetooth, which has a minimum effect (Ghayvat et al. 2014, 2015d, 2016).

Figure 4.28 shows the level of degradation caused by different factors. The RSSI value has been majorly affected by S_{IS-Rx}, followed by material attenuation, by S_{Tx-Rx}, and then by DOA. The selection of channel has least affected the RSSI value. Figure 4.29 presents the effect of different factors on packet reliability metric PSR. Similar to RSSI, the PSR metric is severely affected by the S_{IS-Rx}, followed by S_{Tx-Rx}, S_{Tx-Rx}, and finally, channel selection. The DOA has affected the PSR the least.

Primarily, the IS signal was applied at 2400 MHz frequency, and the corresponding SNR value for Wellness network was measured. However, as IS signal operating frequency was moving close to Wellness network operating frequency, the SNR value of the smart building was degrading. Different sets of experiments had been performed to various interference sources. The frequency offset value, which has been noted for microwave oven, was ±15 MHz, the microwave was the

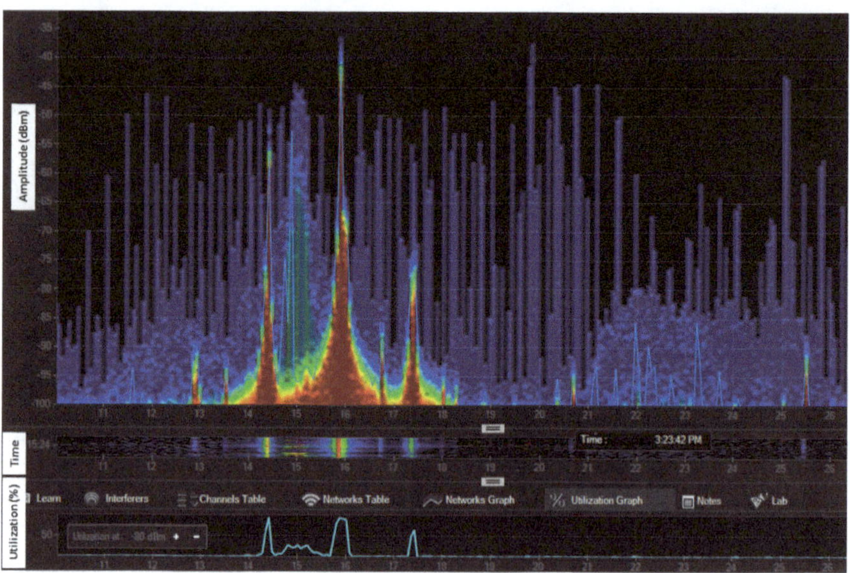

Fig. 4.23 Shows the Wi-Fi functioning over the same frequency 2.430 MHz, which degraded the XBee RF link quality

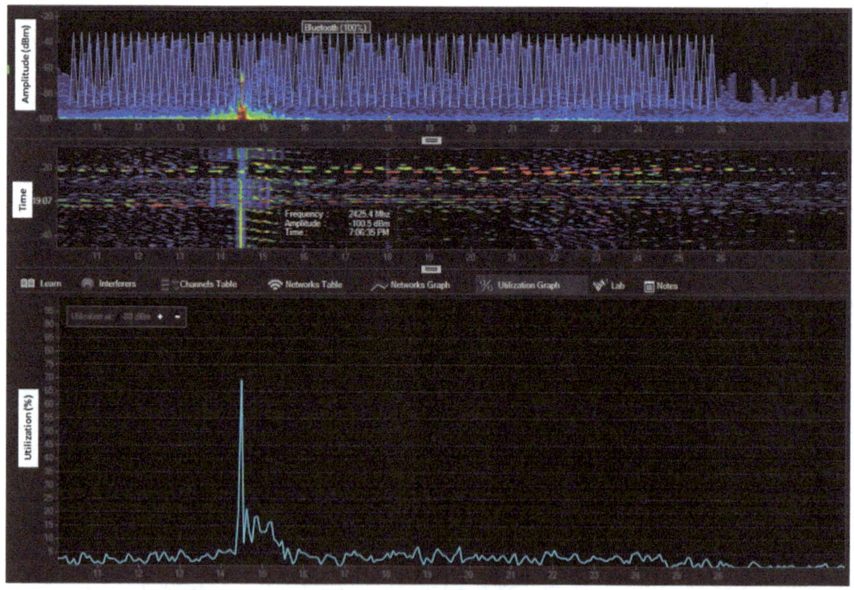

Fig. 4.24 Shows the Bluetooth functioning over the same frequency 2.430 MHz, which degraded the XBee RF link quality

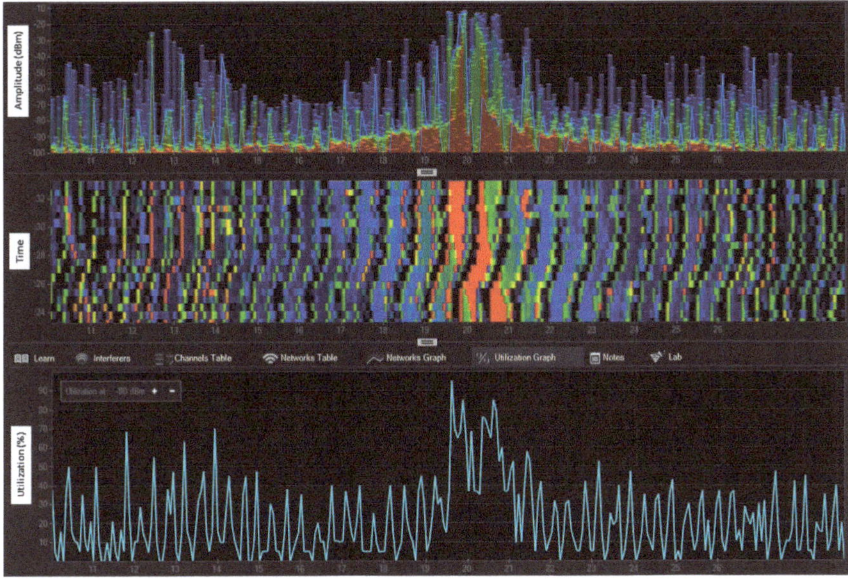

Fig. 4.25 Microwave oven distributed all ZigBee channels, and the microwave signal are dissipated across the whole ZigBee spectrum

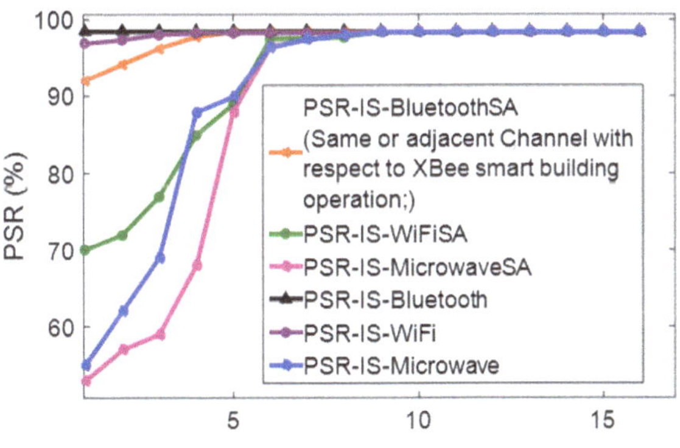

Fig. 4.26 PSR as a function of the distance between IS and Rx, S_{IS-Rx} = 1 to 16 m

strongest source of interference, and it had recorded highest offset value. Followed by WiFi, for WiFi, the frequency offset was ±11 MHz. Whereas Bluetooth had recorded least offset value just ±4 MHz. The whole setup for frequency offset measurement is shown in Fig. 4.30.

Every time for the particular deployment the designer has to perform the measurement and analysis as illustrated above.

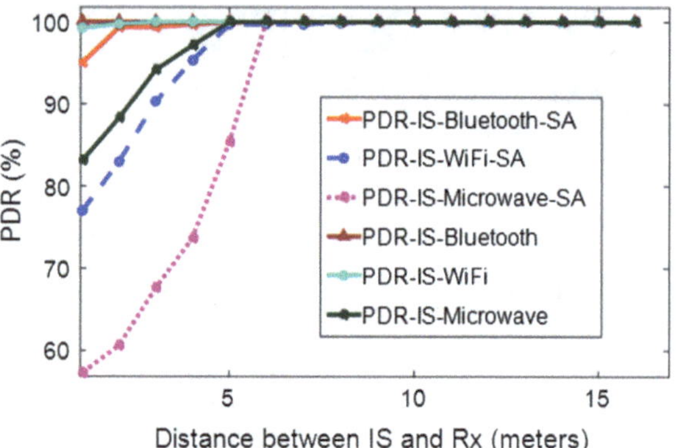

Fig. 4.27 PDR as a function of distance between Tx and Rx, S_{IS-Rx} = 1 to 16 m

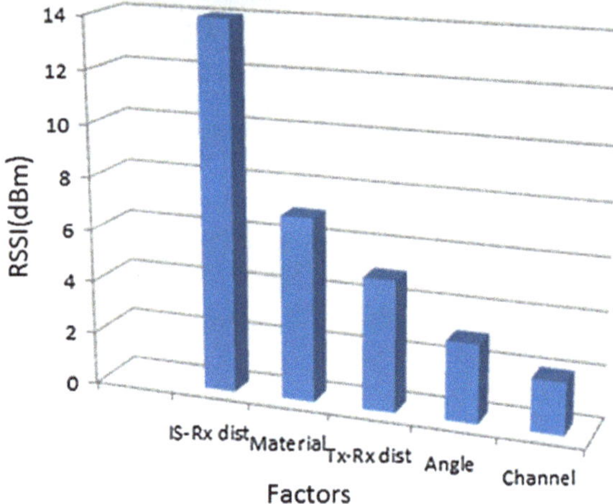

Fig. 4.28 Difference in average RSSI between the highest and lowest levels of each factor

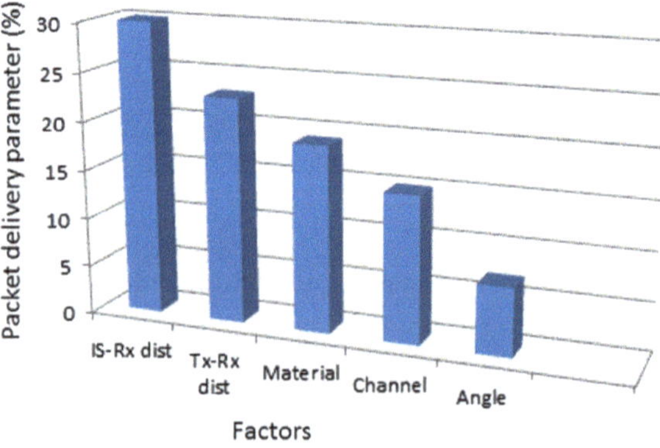

Fig. 4.29 Difference in average packet delivery parameter between the maximum and minimum levels for each factor

Fig. 4.30 Offset frequency measurement by RF spectrometer

4.5 Conclusion

The approach is to coexist with these 2.4-GHz technologies, tolerating these interferences and building material without causing the disproportionate degradation. The change in the standard architecture and functioning principle of other technologies is not feasible. Additionally, the change in building materials such as a door, walls, and other household stuff are not allowed. The question is how designers can get the optimum performance under these constraint operating environments? A designer can control the deployment approach and implementation of Wellness protocol. Furthermore, there are procedures to mitigate the losses. The present research work helps researchers to find the suitable deployment approach for optimum output in the real world condition.

References

Ghayvat, H., Mukhopadhyay, S., and Gui, X.: 'Issues and mitigation of interference, attenuation and direction of arrival in IEEE 802.15. 4/ZigBee to wireless sensors and networks based smart building', Elsevier: *Measurement,* 2016, 86, pp. 209–226.

Metageek, (2016). *Metageek device.* Retrieved Nov 19, 2016, from http://www.metageek.com/index-2.html?utm_expid=190328-155.9nhw_yWyRJK34laAPe_qUw.1&utm_referrer=https%3A%2F%2Fwww.google.co.nz.

H. Ghayvat, S. Mukhopadhyay, X. Gui, and N. Suryadevara, "WSN-and IOT-Based Smart Homes and Their Extension to Smart Buildings," MDPI: *Sensors,* vol. 15, pp. 10350–10379, 2015c.

Ghayvat, H., Liu, J., Alahi, M., Mukhopadhyay, S., and Gui, X.: 'Internet of Things for smart homes and buildings: Opportunities and Challenges', *Australian Journal of Telecommunications and the Digital Economy*, 2015a, 3, (4), pp. 33–47

H. Ghayvat, A. Nag, N. Suryadevara, S. Mukhopadhyay, X. Gui, and J. Liu, "SHARING RESEARCH EXPERIENCES OF WSN BASED SMART HOME," *International Journal on Smart Sensing & Intelligent Systems,* vol. 7, 2014, pp. 1997–2013.

H. Ghayvat, S. C. Mukhopadhyay, and X. Gui, "Sensing Technologies for Intelligent Environments: A Review," in *Intelligent Environmental Sensing,* ed: Springer, 2015d, pp. 1–31.

J. Liu H. Ghayvat, and S. Mukhopadhyay, "Introducing Intel Galileo as a development platform of smart sensor: Evolution, opportunities and challenges," *in Industrial Electronics and Applications (ICIEA),* 2015 IEEE 10th Conference on, Auckland NZ, 15–17 June 2015, pp. 1797–1802.

Ghayvat, H., Mukhopadhyay, S., Liu, J., and Gui, X.: 'Wellness Sensors Networks: A Proposal and Implementation for Smart Home to Assisted Living' *IEEE Sensors Journal,* 2015b, Volume: 15, Issue: 12, pp. 7341–7348.

Chapter 5
Activity Detection and Wellness Pattern Generation

Abstract This chapter presents a novel near real-time sensor segmentation approach that incorporates the notions of the sensor, location, and time correlation. The major objectives and contribution of this research study are: dynamic sensor event segmentation for real-time activity recognition; dataset collection and variety; and Machine Learning algorithms. The rest of the chapter includes the classification of activities, development of wellness belief and wellness function to pattern generation. In the end, the web-based results of wellness system have been shown.

5.1 Introduction

The ambient assisted living (AAL) is a technology aided environment in which an occupant lives independently. To assist in a well-timed and proactive manner, it is inevitable to know the daily activity and events that an occupant performs.

Alt of data from sensor events are collected to represent those activities. Data mining and machine learning techniques have been applied to facilitate interpretation and implication of such data. This research work presents a novel near real-time sensor segmentation approach that incorporates the notions of the sensor, location, and time correlation (Ghayvat et al. 2014; 2015a, b; 2016).

The major objectives and contribution of this research study are as follows:

- **Dynamic sensor event segmentation for real-time activity recognition**: Most of the studies in this area is planned to recognize activities based upon offline and pre-segmented sensor datasets. Whereas real-time activity recognition based on streaming sensor data remains challenging and unsolved (Cook et al. 2013a, b). This study presents a novel dynamic online sensor events segmentation approach, which enable real-time activity recognition and timely forecasting of future activities.
- **Dataset collection and variety**: Majority of research work include laboratory based one or two home datasets. In this research, a real-world smart home

H. Ghayvat and S.C. Mukhopadhyay, *Wellness Protocol for Smart Homes*,
Smart Sensors, Measurement and Instrumentation 24,
DOI 10.1007/978-3-319-52048-3_5

system is designed and deployed in four different houses for long-term monitoring.

- **Machine Learning algorithms**: There are numerous learning algorithms available for activity recognition and forecasting. The performance of these algorithms varies according to datasets. In the present work the wellness algorithm is used and compared with other existing approaches.

The rest of the chapter includes the classification of activities, development of wellness belief and wellness function to pattern generation. In the end, the web-based results of wellness system have been shown.

5.2 Classification of Events and Activities

The activities of daily livings (ADLs) are defined as the record of all routine works that an occupant executes in everyday life without taking any support. The ADLs generally fall into following classes; that is eating, cooking, bathing, dressing, toileting, functional mobility, entertainment (Television) and grooming/hygiene. In AAL environment, the ability of an occupant to perform ADLs is diagnosed according to healthcare perspective. This diagnoses the caring plan on the basis of present potential and health condition. The recognition of activities is done with a high degree of precision using a suitably designed sensing unit. Those ADLs include cooking (for breakfast, lunch and dinner), preparing general food items, making milk/tea/coffee, cleaning dishes by hand, having a meal (breakfast, lunch and dinner), relaxing in a room, watching TV, using toilet (washing hands), showering, going to bed, getting up, waking up during sleep, leaving home and returning home. Besides, province acquaintance plays a decisive role in categorizing and recognizing ADLs. For example in a general case, a house or apartment contains one or more bedrooms, common rooms, bathrooms or entrance areas and car park. Also, some activities can take place only in certain areas of the home, such as an occupant cooks only in the kitchen and clean dishes in the wash basin. Figure 5.1 illustrates a room-level-based ADL representation (Ghayvat et al. 2015).

Human activities are exceptionally dissimilar and complex. So, combining the heterogeneity of lifestyles, practices, and capabilities to form the ADLs is difficult. Consequently, the pattern of performing the daily activities varies from one occupant to another. In the data driven approach of AAL, feature extraction is one of the inevitable steps for activity recognition, which extract the information from raw sensor data. Such information comprises the location, time, duration of sensor activation, environmental context, object usage and movement of the occupant. ADLs are series of sub-activities, for instance, preparing rice is comprised of taking rice in the cooker, washing it in the wash basin and turning on the rice cooker. An orthogonal representation of modeling activities is shown in Fig. 5.2 (Bakar et al. 2016).

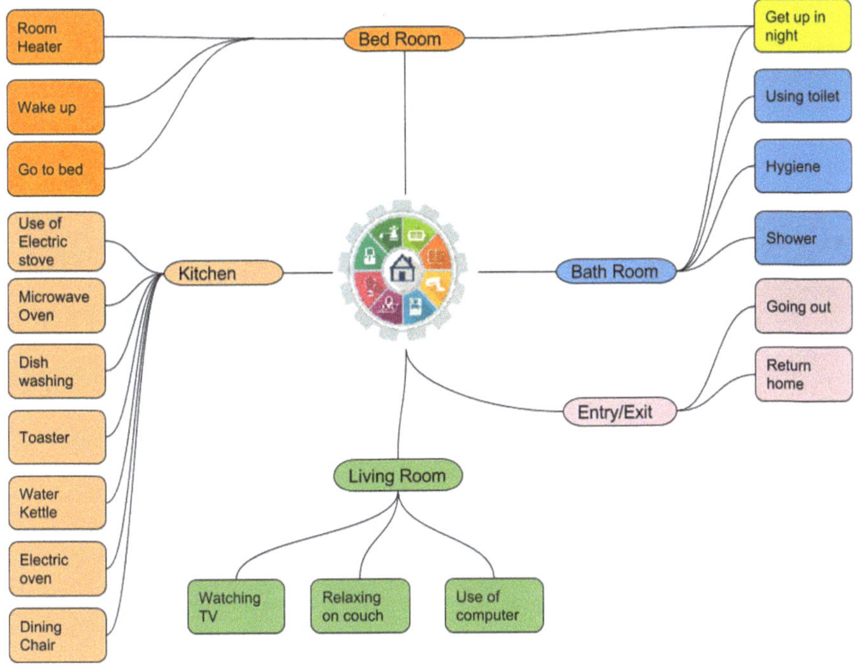

Fig. 5.1 Activities of daily living

The ADLs recognition system design comprises three levels to extract the information from raw sensor data. These levels are as follows:

Sensor Activation Level (**1**): This level contains all kinds of sensor activation due to events performed by the occupant. This data is a unitless number which does not have any decision-making information in the current condition. This data is sent to the upper level for contextual recognition.

Contextual Parameter extraction Level (**2**): This level recognizes the basic ADLs on the basis of location, time and context. This basic ADLs are sent to upper-level recognition.

ADLs discovery level (**3**): The basic ADLs are labeled and correlated depending on contextual information for pattern generation.

To authenticate the activity recognition algorithms, the experiment was conducted in four houses. The houses were equipped with heterogeneous sensing units. Additionally, one wearable accelerometer has been attached to the occupant. Table 5.1 illustrates an overview of the sensor selection and deployment in the smart home environment.

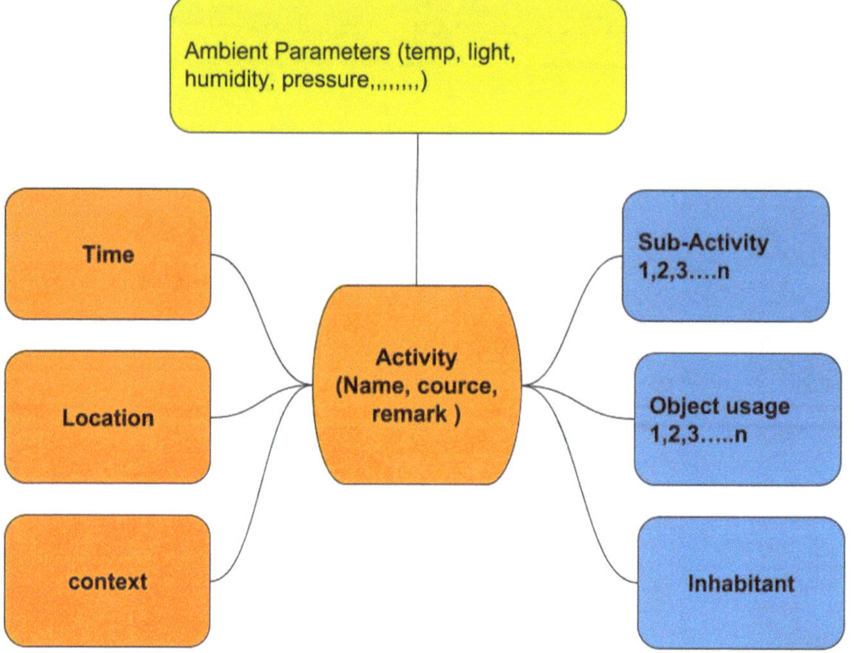

Fig. 5.2 Modeling sub-activity for ADLs

5.3 Activity Annotation

Sensor events are usually logged in the receiver computer continuously over time. The recorded data related to sensor activations includes the start and end time stamp of the activation along with the sensor ID number. Sensor location information can be found either directly or indirectly through sensor allocation layout and ID. Figure 5.3 provides an example of a sensor data set.

An activity of an occupant can be observed as a subset of a more abstract activity. For example, preparing a meal may be preparing breakfast, preparing lunch, preparing dinner and preparing tea. For the ADLs annotation, the hierarchical structure is designed. This structure decomposes the ADLs into different models of sub-activities. The lowest tier of the hierarchy consists of the components responsible for events in the home environment. The second level is for context and task identification. This identification is defined as the lowest level of sub-activity abstraction. The highest level of the tier is the ADLs discovery by series of sub-activities from the lowest tier.

Some activities are only performed at the certain time of the day. For example, preparing breakfast takes place only in the morning. There are two kinds of activities interleaved, and non-interleaved. The interleaved activity is incorporated into another activity builds a complex activity flow. For example, an inhabitant may

Table 5.1 Selection and deployment location of sensing units

Activities	Sensor type	Installation location
Prepare food	PIR, light, temperature	Kitchen desk
	Contact	Cabinets, fridge doors, kitchen door
	Gas, smoke	Kitchen ceiling
	E and E unit	Plugged with rice cooker, oven, toaster, water kettle, microwave oven
Washing dishes	PIR	Kitchen window, appliances
	Contact	Cabinets
Having meal	PIR	Kitchen corner
	Pressure	Dining chair
Watching TV	PIR	Living area
	Pressure	Chairs, sofa
	E and E unit	TV plug
Toilet	PIR	Corner of bathroom
	Pressure	Toilet seat
Shower	Contact	Shower door
Get up	PIR	Corner of bedroom
	Pressure	Below the bed
Go to bed	PIR	Corner of bedroom
	Pressure	Below the bed
All indoor activities	Accelerometer	Neck
Relax	Pressure	Chair

use the laptop to read the cooking recipe and wash hands while preparing food to touch the laptop. While non-interleaved activity is simple, such as while using toilet an occupant can not sleep in the bedroom. According to the above discussion, the majority of sensors and objects are kept physically fixed at one place, so that sensor activation information can be related to user's location context.

The activities are the interaction of an inhabitant with the household objects. The monitoring of those activities can be achieved by analyzing the activation of sensing units. These sensing units are connected to the objects. In Fig. 5.4, the sensor activations and object usage are related to activity formation.

All the sensor data is logged into local home gateway server. The activity annotation is done from the collected data. Table 5.2 shows the activity annotation according to the time duration, sensor ID, and time of object usage. In the table below, the experiment is done with different time duration and at a different time of the day.

The next section presents the methodology for anomaly detection based on wellness approach.

+ Options

←T→				Device_ID	SID_DT	Channel_no	Value
☐	Edit	Copy	Delete	407B6A07	2015-09-25 09:10:34	0	137
☐	Edit	Copy	Delete	407B6A07	2015-09-25 09:10:34	1	141
☐	Edit	Copy	Delete	407B6A07	2015-09-25 09:10:34	2	135
☐	Edit	Copy	Delete	407B6A07	2015-10-02 08:24:45	0	167
☐	Edit	Copy	Delete	407B6A07	2015-10-02 08:24:46	1	171
☐	Edit	Copy	Delete	407B6A07	2015-10-02 08:24:46	2	165
☐	Edit	Copy	Delete	407B6A07	2015-10-02 08:24:59	0	167
☐	Edit	Copy	Delete	407B6A07	2015-10-02 08:24:59	1	171
☐	Edit	Copy	Delete	407B6A07	2015-10-02 08:24:59	2	167
☐	Edit	Copy	Delete	407B6A07	2015-10-05 18:34:53	0	155
☐	Edit	Copy	Delete	407B6A07	2015-10-05 18:34:53	1	159
☐	Edit	Copy	Delete	407B6A07	2015-10-05 18:34:53	2	153
☐	Edit	Copy	Delete	407B6A07	2015-10-05 18:34:56	0	155
☐	Edit	Copy	Delete	407B6A07	2015-10-05 18:34:56	1	159
☐	Edit	Copy	Delete	407B6A07	2015-10-05 18:34:56	2	153
☐	Edit	Copy	Delete	407B6A07	2015-10-06 07:40:31	0	149
☐	Edit	Copy	Delete	407B6A07	2015-10-06 07:40:31	1	153
☐	Edit	Copy	Delete	407B6A07	2015-10-06 07:40:31	2	149
☐	Edit	Copy	Delete	407B6A07	2015-10-25 23:39:48	0	157
☐	Edit	Copy	Delete	407B6A07	2015-10-25 23:39:48	1	159
☐	Edit	Copy	Delete	407B6A07	2015-10-25 23:39:48	2	155
☐	Edit	Copy	Delete	407B6A07	2015-10-27 18:09:15	0	147

Fig. 5.3 Sample of Sensor activation logged

5.4 Wellness Belief Model

The present section includes the wellness probabilistic model based behavior analysis approach for ambient assisted living. The motivation is to monitor and track an occupant's wellness by recognizing the malfunctioning and noise received from sensor data. The generation of behavior pattern and an anomaly in routine is discovered through the noise and error corrected received data. Mostly the warning of panic or emergency is related to any wellness risk. This risk is the function of

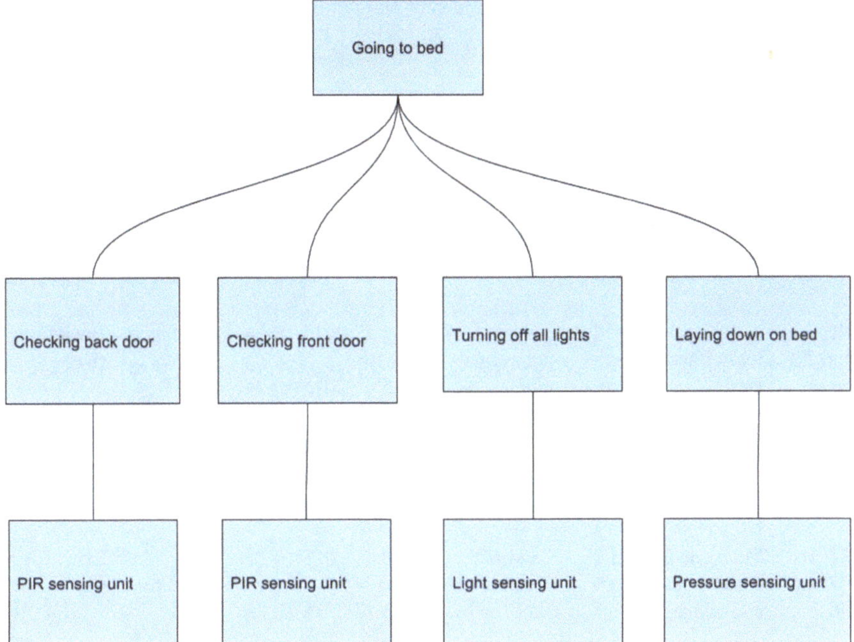

Fig. 5.4 Representation of sensor activation, object usage, and activity

compelling deviation in the daily activities of the inhabitant. Sometimes the smart home system misjudges the distorted and faulty sensor data as variation in the behavior of an occupant and generates the false alarm. The false alarm is the critical issue of pattern generation and forecasting in the smart home system.

5.4.1 Methodology

The activities which the wellness system considers for the pattern generation are sleeping, cooking, eating, toilet, hygiene, relaxing and watching TV. The model is simple and does not demand too much training data. The heterogeneous sensors are deployed into the uncontrolled home to monitor kitchen, living room, bathroom and common room. These sensors are embedded sensing units to record the on/off status of electronics appliances and ambient parameters (temperature, pressure, and movement). The urban indoor communication environment is full of interferences, path and attenuation losses. These interferences introduce the noise and cause the distortion in the sensing data. Sometimes the equipment failure due to wear and tear, loose plug into power supply also show the mysterious behavior. The probabilistic and numeric wellness belief figure is attributed to receiving sensor data to

Table 5.2 Activity annotation process at different portion of the day

Sensor ID	Object connected	Sensor used	Timestamp	Duration of use	Activity annotation
P1	Bed	Pressure	22:04:23 2015-10-23 Start SL 07:33:22 2015-10-24 End SL	9 h 37 min	Sleeping (SL)
P2	Toilet seat	Pressure	07:37:12 2015-10-24 Start TL 07:48:10 2015-10-24 End TL	11 min	Toilet (TL)
E1, E2 and E3	Microwave oven, toaster and water kettle	E and E monitoring unit	07:55:18 2015-10-24 Start BF 08:15:03 2015-10-24 End BF	20 min	Breakfast (BF)
C1	Shower door	Contact	09:40:20 2015-10-24 Start SW 09:58:13 2015-10-24 End SW	18 min	Shower (SW)
E1, E4, E5 and E6	Microwave, electric stove, rice cooker and fridge	E and E monitoring unit	10:30:43 2015-10-24 Start LN 11:53:57 2015-10-24 End LN	1 h 23 min	Lunch (LN)
P5	Dining chair	Pressure	12:30:26 2015-10-24 Start HL 13:11:07 2015-10-23 End HN	41 min	Having lunch (HL)

demonstrate the level of belief in the model. The wellness belief model is modified time to time as new training data gained.

To identify the belief figure in received sensing data, the analysis is done by wellness belief model. If the belief figure is greater than or equal to threshold value then data is forwarded for high-level activity detection and forecasting, otherwise discarded to further analysis when the mysterious behavior of sensing is detected.

5.4.2 *Modeling the Wellness Belief*

Most of the hardware devices get affected by noise which can be calibrated while system design. However, the equipment failure such as producing constant or stationary behavior needs to be identified. The sensing unit produces two types of values, first is continuous (analog) and second is binary (discrete or digital). In both types of values due to the atypical (faulty) behavior of sensing unit either they cause under fitting (fewer data) or overfitting (excess data). For the analog value, such as temperature from the received observations the value should be within defined range. If it goes beyond allowable value or offers constant-stationary value, then there is a high degree of belief that shows device atypical behavior. The wellness thresholding for analog value based sensing units is implemented through the basic linear regression approach. At any time instance t, the wellness belief figure (F_t) is described through the distance (D_t) between the modeled value (M_t) and the actual value (A_t) on standard deviation (σ) and κ is confidence level (0.95). The modeled value function is given by $m(t)$.

$$M_t = \frac{1}{N}\sum_{n=0}^{N} m(t) \tag{5.1}$$

$$F_t = \frac{\sigma}{|D_t|} \quad \text{when } D_t > \sigma \tag{5.2}$$

Otherwise

$$F_t = 0.99 \quad \text{when } D_t \leq \sigma \tag{5.3}$$

Binary values or discrete valued sensing events are modeled through the Poisson distribution. The wellness belief value by Poisson distribution defines the probability of occurrence of an event in the fixed time frame. Suppose that some independent event to occur 'λ' times over a specified time interval, then the probability of exactly 'x' occurrences is equal to

$$F(x, \lambda) = \frac{\lambda^x e^{-\lambda}}{x!} \tag{5.4}$$

The model-based methodology is applied for the recognition of deviation from the anticipated. The forecasted output is compared with actual output and the difference between them is defined as residue. The events are modeled by the probability distributions of parameters; place, time, and units. For efficient detection, even the smallest residue must be determined. Therefore, the observation sampling rate must be satisfactorily high enough for the recognition of the lowest residue. By Nyquist sampling theorem, the sampling rate needs to be at least twice the (peak) rate of variation of sensing outputs. The heterogeneous sensing units are deployed into the home, and sensor status has been recorded as a time series of sampled events. The home conditions are uncontrolled, so various sources are going to add noise such as RF home signals from household objects. Sometimes the sensor event occurrence is shorter than the configured sampling time during data dispensation. This introduces noise into the data signal.

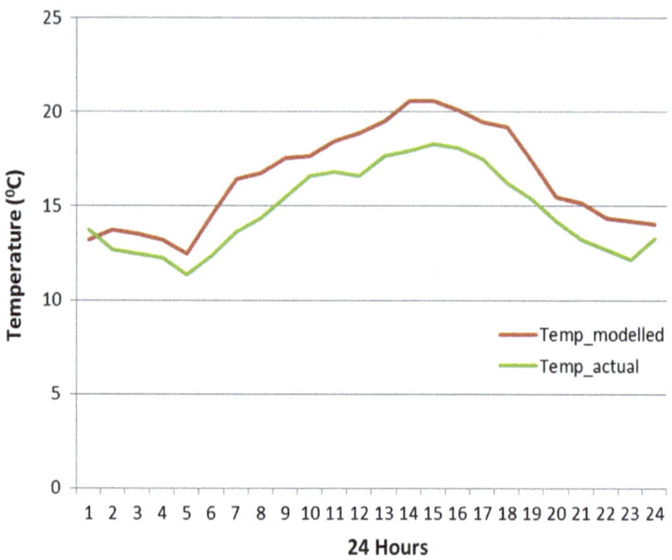

Fig. 5.5 Smart home room temperature actual and expected for 24 h

The sensing device modeling is divided into two sections; analog and discrete.

a. **Analog output sensing unit**: The analog output sensing units such as temperature was modeled using linear regression. The regression function was revised daily for accumulative daily mean, as shown in Fig. 5.5. A_t shows the actual temperature observations, while M_t shows a modeled predicted value. If the difference between them is within the standard deviation, then the belief is set to 0.99. Belief less than 0.77 is labeled as a potential failure. Even with low wellness belief figure, the system can generate behavioral pattern detection and prediction analysis, but the result will be flagged with low belief caution.
b. **Discrete output sensing unit**: The discrete valued device model such as for movement monitoring unit based on 0/1 status. To predict wellness belief value, we use the Poisson probability mass function. For instance, the average rate of opening and closing the door in between 7 AM to 11 AM is 3.4, so the probability of 6 occurrences of the same event is given by 0.072. The probability is very low; it shows an anomaly in sensors or behavior of an occupant. Figure 5.6 shows the Probability of movement in different locations of home based on PIR sensing unit, while Fig. 5.7, presents the probability of occupancy in various places of home based on Force sensing unit.

The wellness belief model is accurate to detect the equipment fault but not full proof for behavioral forecasting and detection, so the wellness system data further analyzed through wellness index.

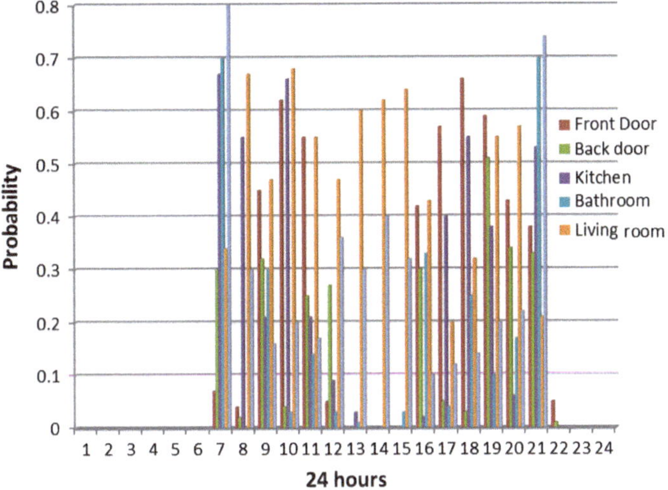

Fig. 5.6 Probability of movement in different locations of home based on PIR sensing unit

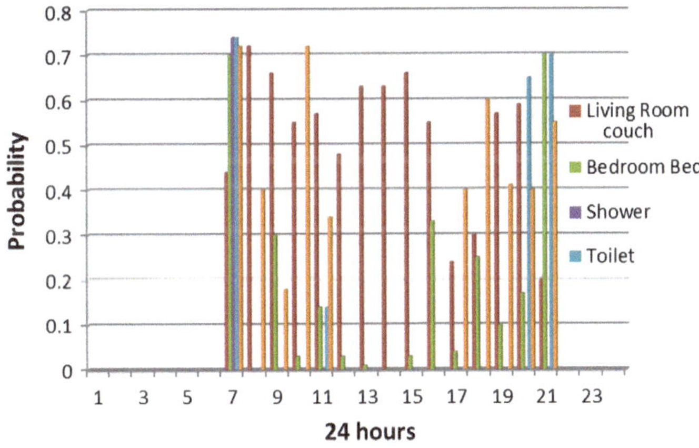

Fig. 5.7 Probability of occupancy in different locations of home based on force sensing unit

5.5 Wellness Determination of an Occupant

The caregiver, as well as healthcare professionals, can only assist if they get fully extracted information instead of the time of activities or object usage. The mathematical values are one of the best ways to represent the well-being of an individual. For this task, wellness function was designed. The β_1 and β_2 were two wellness parameters in wellness model. The calculation of β_1 was done through the inactive or non-usage duration of objects while β_2 was measured by active interaction duration of objects. If β_1 and β_2 are equal to 1, then the occupant is healthy.

Initially, Wellness function $\beta_{1,old}$ and $\beta_{2,old}$ were formulated. Later, these old wellness functions have modified to minimize the issue of false alarm.

5.5.1 Old Wellness Function

$\beta_{1,old}$ wellness function was determined through the inactive time period of object usage by the occupant; t is the actual time period of inactive duration of all or particular object, when they were not in use; T_{IN} is the maximum inactive time period during which no object was used and could cause anomaly condition.

$$\beta_{1,old} = 1 - \frac{t}{T_{IN}} \tag{5.5}$$

$$\beta_{2,old} = 1 + \left(1 - \frac{T_A}{T_N}\right) \tag{5.6}$$

where $\beta_{2,old}$ wellness function was determined through the active time period of object usage by the occupant. T_A is the actual time period of active duration of a particular object when they were in use. T_N is the maximum active time period during which object was used and beyond this time duration, it may cause anomaly condition. If T_A is less than or equal to T_N, then no anomaly has to be calculated.

5.5.2 Modified Wellness Function

The issue with old wellness function was, it did not include the seasonal variations and that causes false alarm messages. In new wellness function, the seasonal changes have been introduced through the time series approach. The present approach is dynamic approach, where the wellness

$$\beta_{1,new} = e^{-t/T_{IN}} \tag{5.7}$$

where $\beta_{1,new}$ wellness function is determined through the inactive time period of object usage by the occupant; t is the actual time period of inactive duration of all or particular object, when they were not in use; T_{IN} is the maximum inactive time period during which no object was used and could cause anomaly condition.

$$\beta_{2,new} = e^{\frac{(T_N - T_A)}{T_n}} \tag{5.8}$$

where $\beta_{2,new}$ wellness function is determined through the excessive active time period of object usage by subject; T_A is the actual time period of active duration of a

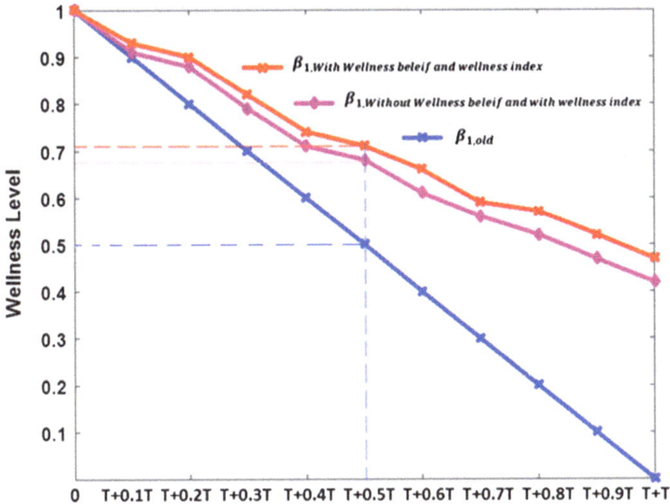

Fig. 5.8 Comparison of $\beta_{1,old}$ and $\beta_{1,new}$ wellness functions (with two cases with wellness belief and without wellness belief)

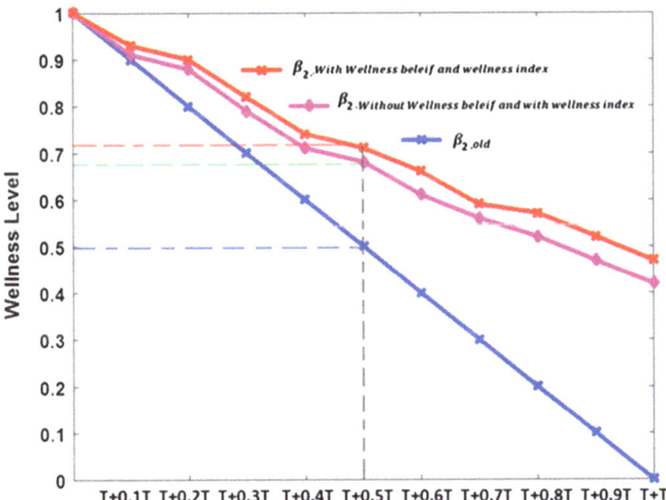

Fig. 5.9 Comparison of $\beta_{2,old}$ and $\beta_{2,new}$ wellness functions (with two cases with wellness belief and without wellness belief)

particular object, when they were in use. T_N is the maximum active time period during which object was used, and beyond this time duration, it may cause anomaly condition.

Figures 5.8 and 5.9 depict the advantage of the modified wellness index β_1 and β_2 respectively on the ground of selected time duration to generate a positive alarm of anomaly condition. Both the figures look almost similar, but they have different significance in the analysis of wellness parameters. In old wellness function for the 50% duration of the time period, the wellness indices had recorded 0.5. Whereas for the same, the new wellness indices have shown improvement and touched 0.68 for without wellness belief and 0.72 with Wellness belief model.

5.5.3 Measurement of Maximum Active and Inactive Object Usage Duration

The routine period of object usage indicates that the person is performing daily activities in a healthy manner. According to seasonal changes, the occupant's behavior and the time duration of object usage changes, this is not an anomaly condition. It should not be indicated as warning alert. For example, an occupant's lifestyle is not going to remain same in winter and summer; there will be a significant change in the activities. That change needs to be identified and separated from anomaly indication.

For the accurate measurement of lifestyle change and anomaly detection, the maximum inactive (T_{IN}) and active (T_N) time of object usage were formulated with respect to days, weeks, months of the calendar and seasonal changes. Time series techniques have been used to derive the model for T_{IN} and T_N.

Definition of symbol used in formulas from Eqs. (5.9) to (5.20):

L number of period in one cycle or seasonal length (for a week 1–7)
α Data smoothing factor 0–1
β Trend smoothing factor 0–1
γ Seasonal change smoothing factor 0–1
x Observation of object at current time
S Smoothed observation
T_t Trend factor
C Seasonal Trend/Index
F The forecast at "m" periods ahead
t Index to show the time period
m number of steps ahead of forecast

The smoothed observation for inactive excessive usage was measured by the recently received sensing events $\{\alpha(x_t - C_{t-L}) + (1-\alpha)(S_{t-1} - T_{INt-1})\}$ and the event occurred during same season in last year $\{\alpha'(x'_t - C'_{t-L}) + (1-\alpha')(S'_{t-1} + T'_{INt-1})\}$.

$$S_t = \frac{1}{2}\left[\{\alpha(x_t - C_{t-L}) + (1-\alpha)(S_{t-1} - T_{INt-1})\} + \left\{\alpha'\left(x'_t - C'_{t-L}\right) + \left(1-\alpha'\right)\left(S'_{t-1} + T'_{INt-1}\right)\right\}\right] \tag{5.9}$$

The inactive excessive usage was measured by the recently received sensing event parameters $\{\beta(S_t - S_{t-1}) + (1-\beta)(T_{INt-1})\}$ and the event occurred during same season in last year $\left\{\beta'\left(S'_t - S'_{t-1}\right) + \left(1-\beta'\right)\left(T'_{INt-1}\right)\right\}$.

$$\mathrm{T_{IN}} = \frac{1}{2}\left[\{\beta(S_t - S_{t-1}) + (1-\beta)(T_{INt-1})\} + \left\{\beta'\left(S'_t - S'_{t-1}\right) + \left(1-\beta'\right)\left(T'_{INt-1}\right)\right\}\right] \tag{5.10}$$

The seasonal trend for inactive excessive usage was measured by the recently received sensing event parameters $\{\gamma(x_t - S_t) + (1-\gamma)(C_{t-L})\}$ and the event occurred during same season in last year $\{\gamma'(x'_t - S'_t) + (1-\gamma')(C'_{t-L})\}$.

$$\mathrm{C_t} = \frac{1}{2}\left[\{\gamma(x_t - S_t) + (1-\gamma)(C_{t-L})\} + \left\{\gamma'\left(x'_t - S'_t\right) + \left(1-\gamma'\right)\left(C'_{t-L}\right)\right\}\right] \tag{5.11}$$

The smoothed observation for active excessive usage was measured by the recently received sensing event parameters $\{\alpha(x_t - C_{Nt-L}) + (1-\alpha)(S_{Nt-1} - T_{Nt-1})\}$ and the event occurred during same season in last year $\{\alpha'(x'_t - C'_{Nt-L}) + (1-\alpha')(S'_{Nt-1} + T'_{Nt-1})\}$.

$$\mathrm{S_{Nt}} = \frac{1}{2}\left[\{\alpha(x_t - C_{Nt-L}) + (1-\alpha)(S_{Nt-1} - T_{Nt-1})\} + \left\{\alpha'\left(x'_t - C'_{Nt-L}\right) + \left(1-\alpha'\right)\left(S'_{Nt-1} + T'_{Nt-1}\right)\right\}\right] \tag{5.12}$$

The active excessive usage was measured by the recently received sensing event parameters $\{\beta(S_{Nt} - S_{Nt-1}) + (1-\beta)(T_{Nt-1})\}$ and the event occurred during same season in last year $\{\beta'(S'_{Nt} - S'_{Nt-1}) + \left(1-\beta'\right)(T'_{Nt-1})$.

$$\mathrm{T_N} = \frac{1}{2}\left[\{\beta(S_{Nt} - S_{Nt-1}) + (1-\beta)(T_{Nt-1})\} + \left\{\beta'\left(S'_{Nt} - S'_{Nt-1}\right) + \left(1-\beta'\right)\left(T'_{Nt-1}\right)\right\}\right] \tag{5.13}$$

The seasonal trend for active excessive usage was measured by the recently received sensing event parameters $\{\gamma(x_t - S_{Nt}) + (1-\gamma)(C_{Nt-L})\}$ and the event occurred during same season in last year $\{\gamma'(x'_t - S'_{Nt}) + (1-\gamma')(C'_{Nt-L})\}$.

$$\mathrm{C_{Nt}} = \frac{1}{2}\left[\{\gamma(x_t - S_{Nt}) + (1-\gamma)(C_{Nt-L})\} + \left\{\gamma'\left(x'_t - S'_{Nt}\right) + \left(1-\gamma'\right)\left(C'_{Nt-L}\right)\right\}\right] \tag{5.14}$$

$$\mathrm{F_{t+m} = S_t + mT_{IN} + C_{t-L+1+((m-1)modL)}} \tag{5.15}$$

$$F_{Nt+m} = S_{Nt} + mT_N + C_{Nt-L+1+((m-1)\mathrm{mod}L)} \tag{5.16}$$

$$S_t = \frac{1}{L}(x_1 + x_2 + x_3 + x_4 \ldots x_L) \tag{5.17}$$

$$S_{Nt} = \frac{1}{L}(x_1 + x_2 + x_3 + x_4 \ldots x_L) \tag{5.18}$$

$$C_1 = x_1 - S_t,\ C_2 = x_2 - S_t \ldots \tag{5.19}$$

$$C_{N1} = x_1 - S_{Nt},\ C_{N2} = x_2 - S_{Nt} \ldots \tag{5.20}$$

$$T_t = \left[\left(\frac{1}{L}\right)\left\{\left(\frac{x_{L+1} - x_1}{L}\right) + \left(\frac{x_{L+2} - x_2}{L}\right) + \left(\frac{x_{L+3} - x_3}{L}\right) + \ldots + \left(\frac{x_{2L} - x_L}{L}\right)\right\}\right] \tag{5.21}$$

5.6 Experimental Analysis, Observation, and Results

The wellness indices (β_1 and β_2) were evaluated by implementing the system in four different houses and analyzing the recorded activities. Figures 5.10, 5.11, 5.12 and 5.13, present the object usage pattern of four different AAL houses. The heterogeneous sensing units are deployed were the home at a different location and connected to the different object to monitor the individual. The object under monitoring are as follows: fridge door (FGD), wash basin (WB), audio device (AD), television (TV), toaster (TTR), water kettle (WK), electric oven (EO), microwave (MO), room heater (RH), toilet (TL), shower (SH), dining chair (DC), bed (BD), front door (FD) and back door (BKD).

In this smart home framework, sensor ID is coded with an identifier of the functioning area. This ID is combined with the sensor correlation table. The smart home functioning areas can be further clustered as a bed, bath, kitchen, dining, living and home entrance. By investigating the long-term user-annotated sensor data for the activities of cooking and food preparation, having meals, washing dishes, 91.04% of the sensor activations occur in the kitchen and dining area. For the activities of watching television, relaxing on Couch, 92.10% sensors are initiated in the living area. Furthermore, 100% of sensor activations for activities of sleep, wake up during sleep, a bed to toilet, shower, and bathing take place in the area of bed and bath. Also, for the activity of coming back home and leaving home, 98.35% sensor activations occur in the home entrance area. As that the occupant retains a moderately consistent routine, it is supposed that there will be no substantial changes to the daily routine. Thus, different threshold parameters of the maximum activity, inactivity period and sensor alert distance time can be calculated according to the functioning areas.

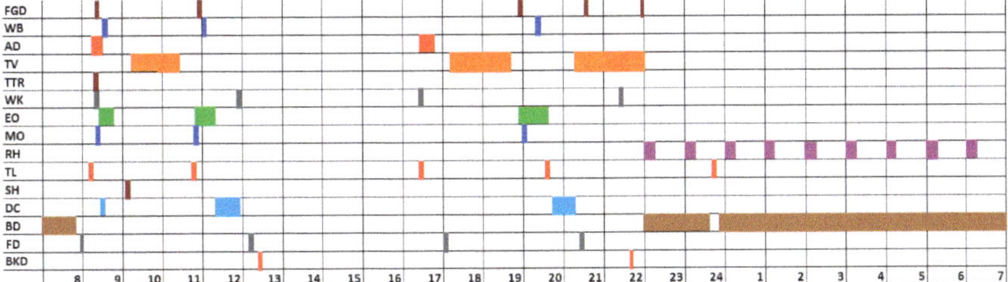

Fig. 5.10 Object usage for one day for house-1

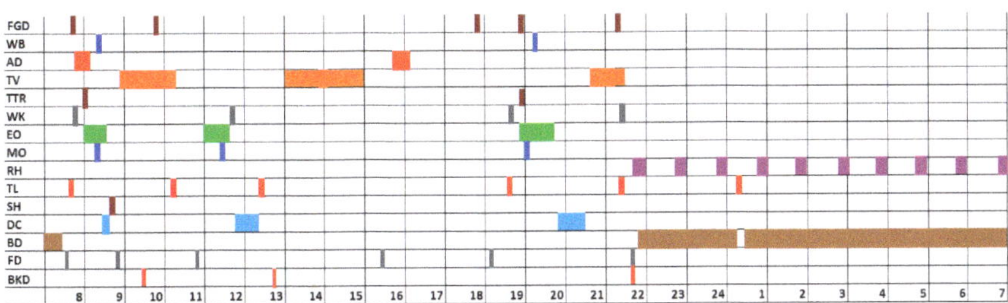

Fig. 5.11 Object usage for one day for house-2

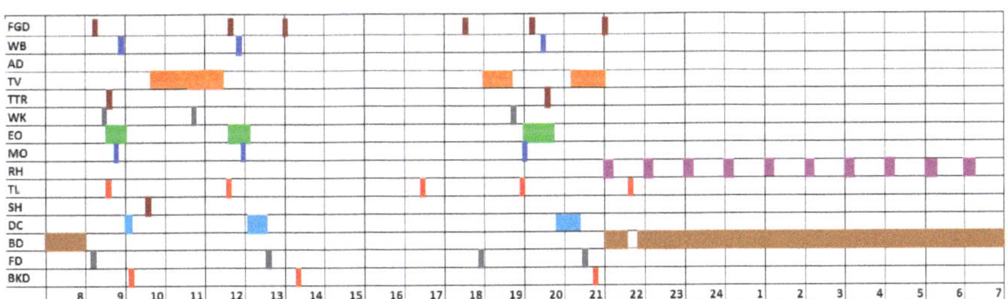

Fig. 5.12 Object usage for one day for house-3

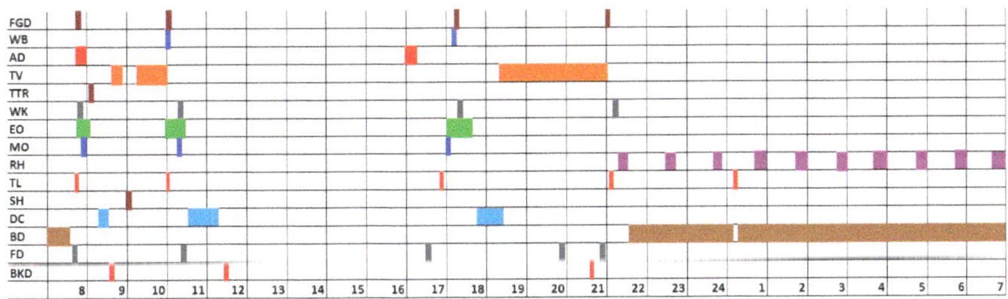

Fig. 5.13 Object usage for one day for house-4

Table 5.3 Number of sensor activation and activity detection for four different houses equipped with hydrogenous sensing units

Title	Home 1	Home 2	Home 3	Home 4
Number of sensors	36	29	38	30
Days monitored	300 days	300 days	300 days	300 days
Sensor activation	701,405	591,359	731,021	648,321
Activity detection	13,290	12,345	14,702	11,257

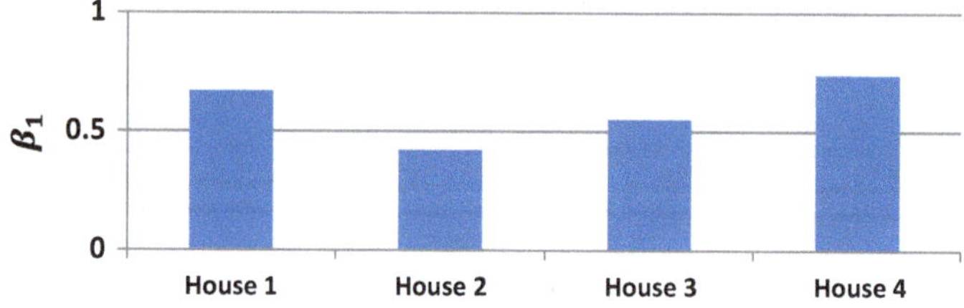

Fig. 5.14 β1,old at four different elderly houses

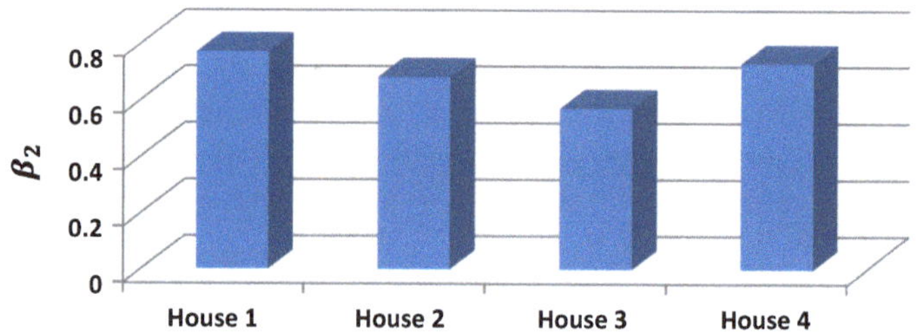

Fig. 5.15 Wellness indices for sleeping activity for four different houses up to one week

Table 5.3, shows the number of sensor activation which is logged into system varies from one subject house to another, and these activations help in determining the activity of daily living.

Figure 5.14, it is detected that the β_1 for the house two on a particular day has gone below 0.5 for the leaving and coming back to home. It is not because of an unhealthy condition. Actually, that day front door was open.

Figure 5.15, presents the β_2 based on object usage in four different houses. It is seen that β_2 for subject house three has gone to very low. It was observed that on that day, the occupant had a long sleep because next day was a holiday. These explanations tell clearly about the wellness determination of the system. The alert can be set depending on values of β_1 and β_2. The wellness alert value may be

Table 5.4 Improved Wellness indices for different activities to four different houses

\`Subject	Sensor ID	ADL	Max-active duration(Sec)	Min-duration (Sec)	Actual duration(Sec)	$\beta_{1,\ old}$	$\beta_{2,\ old*}$	ADL message based on old indices	$\beta_{1,\ new}$	$\beta_{2,\ new*}$	Wellness indices based ADL
1	Bed	Sleeping	32,128	25,712	27,491	0.85	NA	Regular (R)	0.89	NA	R
	Eating	Eating	5532	3829	2143		0.53	**Anomaly (A)**		0.77	**R**
	Toilet	Toilet	1835	1321	1521		NA	Regular		NA	R
	Sofa	Relax	2012	1092	1238			Regular			R
	S-Door	Shower (S)	1627	1170	1532			Regular			R
	TV	Television	3852	2438	2842			Regular			R
2	Bed	Sleeping	30,200	22,245	33,212	0.77	0.54	**Anomaly**	0.83	0.66	**A**
	Eating	Eating	4321	3213	3015		NA	Regular		NA	R
	Toilet	Toilet	1732	1245	2223		0.52	**Anomaly**		0.69	**R**
	Sofa	Relax	1823	1138	1426		NA	Regular		NA	R
	S-Door	Shower	1578	1262	1492			Regular			R
	TV	Television	3647	2745	2984			Regular			R
3	Bed	Sleeping	28,431	20,245	27,212	0.80		Regular	0.89		R
	Eating	Eating	3981	2834	4413		0.57	**Anomaly**		0.79	**R**
	Toilet	Toilet	1838	1341	1530		NA	Regular		NA	R
	Sofa	Relax	1749	1369	1393			Regular			R
	S-Door	Shower	1405	1136	1303			Regular			R
	TV	Television	3729	3021	3213			Regular			R
4	Bed	Sleeping	30,261	27,492	26,492	0.83	NA	Regular	0.91		R
	Eating	Eating	3620	2785	3329			Regular			R
	Toilet	Toilet	1839	1384	2318		0.48	**Anomaly**		0.65	**A**
	Sofa	Relax	1640	1243	1620		NA	Regular		NA	R
	S-Door	Shower	1537	1138	1430			Regular			R
	TV	Television	3124	2481	2647			Regular			R

NA No Anomaly

$\beta_{2,old*}$ and $\beta_{2,new*}$ are calculated only when actual duration is greater than maximum duration different for a different occupant. During alert message generation, first, the sound alarm has been triggered in the house. If an occupant responds to this alarm within predefined safe time and turns off, no alert message will be sent to a caregiver or assisting services. The developed home monitoring sensing systems are implemented at four different homes and still running from May 2014. The reason for choosing four houses is the cost of system design and development.

For the process of realizing the functionalities and the significance of upgraded wellness functions, the 300 days of sensor data had been analyzed which was collected from four subject houses. From house 1, there were 42 warning messages for the methods of old wellness functions. These have included excess usage of bed (10), excess usage of dining chair (7), excess usage of the couch (8) and no usage of the object (17). The reason for many bed alarms was a max inactive duration (no usage) belongs to β_1, the occupant had been to a friend's place and did not turn off the system. Whereas, during the same condition $\beta_{1,new}$ had generated only three warning messages.

The wellness functions ($\beta_{2,old}$) of bed, dining chair, and couch usages had generated 25 messages. Whereas wellness function ($\beta_{2,new}$) was able to restrict the generation of warning messages to bed (4), dining chair (2) and sofa (3). Updated time series values had achieved the reduction in warnings and these time series update had helped in enhancing the threshold limit of new wellness indices.

48.18% improvements over the false positives have been recorded through the newly defined wellness function. The improved wellness function and comparison to impaired wellness function has shown in Table 5.4 for four different houses. In the subject house 1 and 3, old wellness function indicates that the max active duration of eating activities had gone beyond the limit and its anomaly. However according to newly defined $\beta_{2,new}$ that day there was a family visitor, and they were sitting and talking while having a meal. From subject house 2, it shows the anomaly condition by both the wellness function old and new. That time the occupant was unhealthy and having rest. From the same subject house 2, old wellness function shows the excess use of the toilet, whereas the new function had found the use of the toilet by a visitor at home. From subject house 4, it shows that access usage and anomaly condition by new and old wellness function because that day no visitor was there and the occupant had an upset stomach.

5.7 Web-Based Results

An IoT-based smart home system had been developed. The data has been collected and passed through data mining and machine learning algorithms for the decision-making process. The final information was uploaded onto the website. The information through this website was only accessible to an authenticated user via a registered email id and password.

Select date 2015-08-30	Time	Food or Medicine/Supplement
	08:03:42	Food
	08:05:21	Medicine
	11:33:18	Food
	11:38:42	Medicine
	15.32.54	Food
	18:33:42	Food
	18:35:30	Medicine
	21:33:42	Food

Fig. 5.16 The ADLs of having medicine and meal throughout the day

The wellness protocol is equipped with manual push button for panic (emergency), shower, cooking, medicine, eating, sleeping, watching television and toilet. This is a supplementary addition to the system. The wellness pattern generation and forecasting are done through the heterogeneous sensor deployment without manual indication. However, there are few activities that the system sometimes fails to identify correctly such as taking meal or medicine on time. For this kind of activities either system use obtrusive monitoring ways such as camera, wearable sensors or accompany caregiver 24/7, which it is not feasible and economical (Cook et al. 2013a, b). In the present research, the manual indication button data helps to verify the ADLs generated by data received from other heterogeneous sensors.

Figures 5.16 and 5.17 show snapshots from the wellness monitoring website. This website contains the data from the most recent few months. To see the monitoring history from a particular day, the client has to select a day. Figure 5.16 shows the information on the non-electrical appliance usage. It shows sleeping, eating and toilet activities: for example, on 30th August, an occupant slept for 9:10:12 (hr:mn:s). Figure 5.17 presents the monitoring of food and medicine of an occupant. The occupant takes medicine just after breakfast in the morning, and the routine of having medicine is thrice a day.

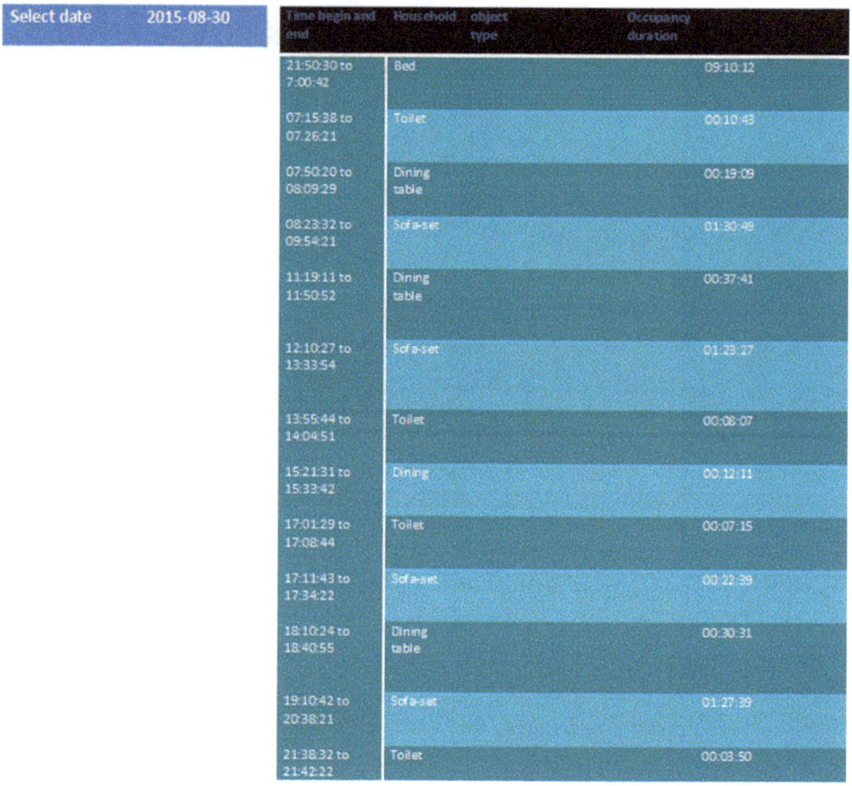

Select date 2015-08-30

Time begin and end	Household object type	Occupancy duration
21:50:30 to 7:00:42	Bed	09:10:12
07:15:38 to 07.26:21	Toilet	00:10:43
07:50:20 to 08:09:29	Dining table	00:19:09
08:23:32 to 09:54:21	Sofa-set	01:30:49
11:19:11 to 11:50:52	Dining table	00:37:41
12:10:27 to 13:33:54	Sofa-set	01:23:27
13:55:44 to 14:04:51	Toilet	00:08:07
15:21:31 to 15:33:42	Dining	00:12:11
17:01:29 to 17:08:44	Toilet	00:07:15
17:11:43 to 17:34:22	Sofa-set	00:22:39
18:10:24 to 18:40:55	Dining table	00:30:31
19:10:42 to 20:38:21	Sofa-set	01:27:39
21:38:32 to 21:42:22	Toilet	00:03:50

Fig. 5.17 The ADLs throughout the day

5.8 Conclusion

The present chapter was designed and implemented a novel integrated framework for wellness detection of the individual. The wellness detection had been done through generating the activity of daily living based on the interaction with different objects in a home environment. This activity detection and wellness pattern generation helps to identify any anomaly change in the lifestyle.

The developed system was based on wellness protocol approach. The system had improved wellness indices β_1 and β_2 which reduced the false positive alarm at best possible level. The system was implemented in four different houses; few of them were very old houses. The system had been designed in such a way that the users did not need to look after its maintenance except small wear and tear, and power supply to sensing system.

The present system was involved in some processes in different stages. It starts from intelligent sensing units which collect the data from the home environment and objects usage. These sensing units are smart enough to analyze the excess and undesired data through intelligent sampling and control algorithm. Ultimately the data received at the server has been analyzed in real time as well as near real-time based on wellness belief model, time series and function. This analysis generates the sufficient information for healthcare assistant and caregiver to monitor and assist the occupant before any unforeseen condition arises.

References

D. J. Cook, N. C. Krishnan, and P. Rashidi, "Activity discovery and activity recognition: A new partnership," *Cybernetics, IEEE Transactions on,* vol. 43, pp. 820–828, 2013a.

D. J. Cook, A. S. Crandall, B. L. Thomas, and N. C. Krishnan, "CASAS: A smart home in a box," *Computer,* vol. 46, 2013b.

Ghayvat, H., Mukhopadhyay, S., and Gui, X.: 'Issues and mitigation of interference, attenuation and direction of arrival in IEEE 802.15. 4/ZigBee to wireless sensors and networks based smart building', Elsevier: *Measurement,* 2016, 86, pp. 209–226.

Ghayvat, H. Mukhopadhyay, S. Liu, J. and Gui, X.: 'Wellness Sensors Networks: A Proposal and Implementation for Smart Home to Assisted Living' *IEEE Sensors Journal,* 2015, Volume: 15, Issue: 12, pp. 7341–7348.

H. Ghayvat, S. Mukhopadhyay, X. Gui, and N. Suryadevara, "WSN-and IOT-Based Smart Homes and Their Extension to Smart Buildings," MDPI: *Sensors,* vol. 15, pp. 10350–10379, 2015.

Ghayvat, H. Liu, J. Alahi, M. Mukhopadhyay, S. and Gui, X.: 'Internet of Things for smart homes and buildings: Opportunities and Challenges', *Australian Journal of Telecommunications and the Digital Economy,* 2015, 3, (4), pp. 33–47.

H. Ghayvat, A. Nag, N. Suryadevara, S. Mukhopadhyay, X. Gui, and J. Liu, "SHARING RESEARCH EXPERIENCES OF WSN BASED SMART HOME," *International Journal on Smart Sensing & Intelligent Systems,* vol. 7, 2014, pp. 1997–2013.

U. Bakar, H. Ghayvat, F. Hasan, and S. Mukhopadhyay, "Activity and anomaly detection in smart home: A survey," *in Next Generation Sensors and Systems,* ed: Springer, 2016, pp. 191–220.

Chapter 6
Wellness Pattern Generation and Forecasting

Abstract The smart home data analysis can be divided into two parts; one, domain is activity recognition that has been discussed in the last chapter, and the other one is wellness pattern generation and forecasting. The forecasting in the WSN based smart home is the dynamic learning from the historical sensing events. Transformation of prior sensing events into pattern and forecast can be done by the analysis of knowledge discovery and soft computing techniques. There are a number of knowledge and soft computing methods available, but these methods do not perform well in the AAL environment. Either these methods are complex and needs large training data or too simple where they offer poor accuracy (Moutacalli et al. 2015; Pulsford et al. 2011; Candás et al. 2014). For the Wellness Protocol based AAL the time series approach has been proposed and implemented. This time series approach includes the seasonal parameters from last year; it does not demand too much learning data. The rest of the chapter includes the wellness forecasting analysis and comparative results with other existing data mining methods.

6.1 Modelling Trends and Forecasting

The evaluation of forecasting model is done by received sensing data. Figure 6.1 shows the general process of the correct assessment of the learning model. The sensing data are separated into two types of sets, training, and test data. Training data produce the model with a Wellness learning algorithm. Once the Wellness model is achieved, the historical test dataset is applied to assess the model (Bakar et al. 2016).

To predict an activity label in real time, a number of steps are identified as follows.

- Step 1
  Measure the basic wellness parameters from recent historical data, seasonal data from last year and calculate wellness indices from the equations derived in chapter five.

H. Ghayvat and S.C. Mukhopadhyay, *Wellness Protocol for Smart Homes*,
Smart Sensors, Measurement and Instrumentation 24,
DOI 10.1007/978-3-319-52048-3_6

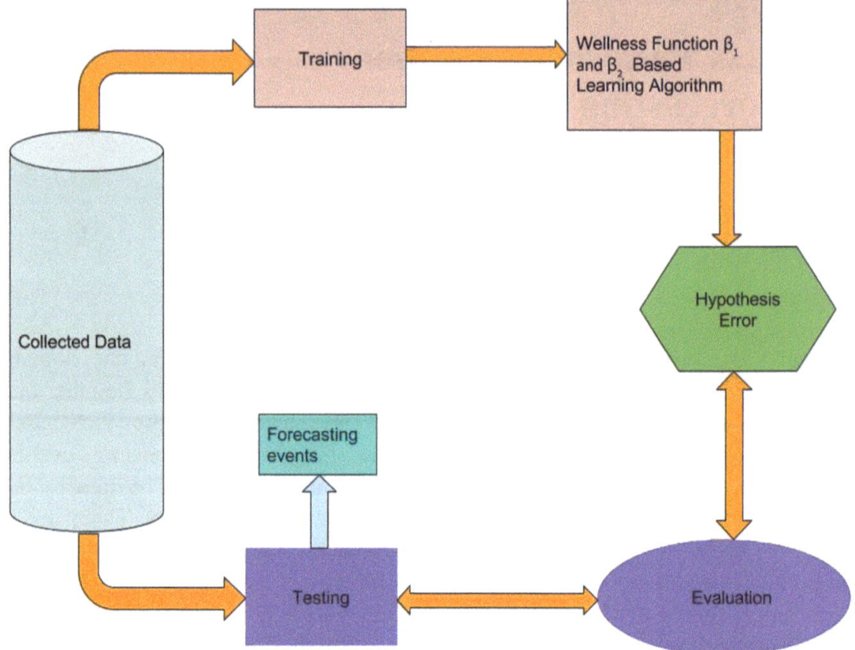

Fig. 6.1 Functional block diagram of Wellness learning algorithm for time series analysis

- Step 2
  Choose one random full-day dataset as the test set, use the rest of the datasets as the training set, reconstruct the model and re-evaluate the model on the test set.
- Step 3
  Construct the activity model utilizing the training set via methodologies that are presented Chap. 5.
- Step 4
  Implement the test set on wellness time series algorithm as online streaming; for each incoming sensor event, apply the sensor segmentation methodology, and use the prebuilt activity model to predict the new activity label.

6.2 Behavioural Pattern Generation and Forecasting

The heterogeneous sensor-based smart home system data from the different subject had been collected for the forecasting process. The forecasting process is computing the most suitably fitted curve from trends and seasonal parameters. In the forecasting process, the priority is given to non-electrical object data usage over electrical object usage data, as electrical appliances get auto programming. Figure 6.2

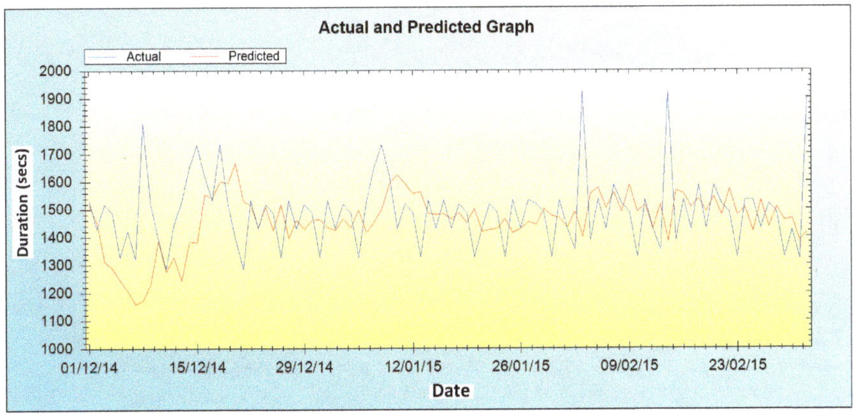

Fig. 6.2 Actual shower usage and its trend

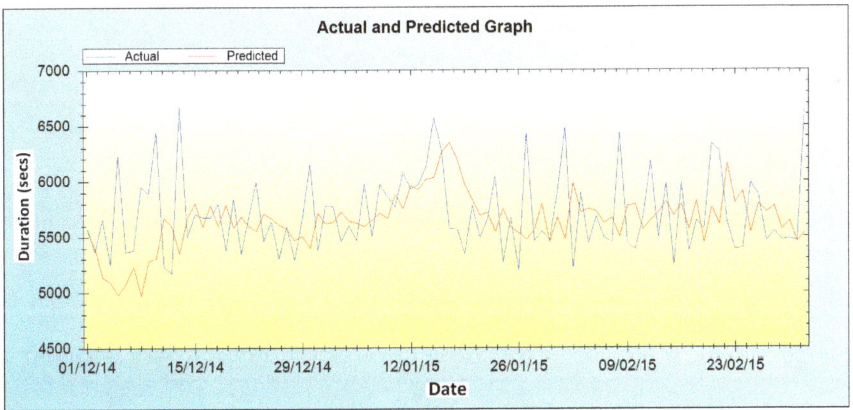

Fig. 6.3 Actual dining chair usage and its trend

shows the shower usage activity duration and its equivalent trend analysis for eleven weeks at an occupant's house living alone (Ghayvat et al. 2015).

Figure 6.3 shows the eating activity (dining chair usage) duration and its equivalent trend analysis for eleven weeks at an inhabitant's house living alone.

Figure 6.4 shows the sleeping activity (bed usage) duration and its equivalent trend analysis for eleven weeks at an inhabitant's house living alone.

Figure 6.5 shows the toilet usage duration and its equivalent trend analysis for eleven weeks at an inhabitant's house living alone.

Based on the data of eleven weeks the data of twelfth and thirteenth weeks has been predicted. It contains the data of recent successive weeks and historical seasonal data from last season to reduce the residue and get close curve fitting. Figure 6.6 depicts the usage of the shower by an occupant for eleven weeks and its corresponding trend, the forecast for the twelfth and thirteenth weeks.

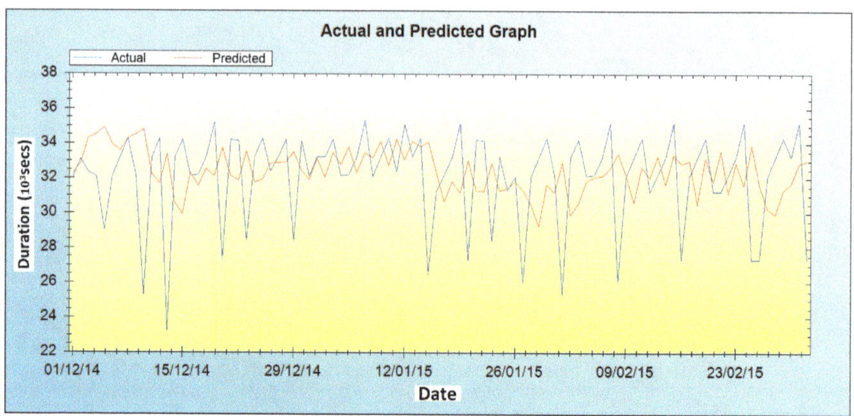

Fig. 6.4 Actual dining bed usage and its trend

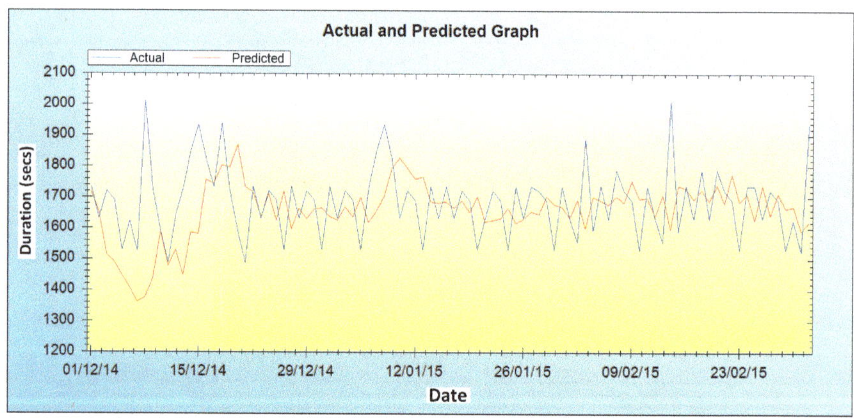

Fig. 6.5 Actual dining toilet usage and its trend

Figure 6.7 depicts the usage of dining chair by an occupant for eleven weeks and its corresponding trend, the forecast for the twelfth and thirteenth weeks.

Figure 6.8 depicts the usage of the bed by an occupant for eleven weeks and its corresponding trend, the forecast for the twelfth and thirteenth weeks.

Figure 6.9 depicts the usage of the toilet by an occupant for eleven weeks and its corresponding trend, the forecast for the twelfth and thirteenth weeks.

By the application of formula derived in chapter five, the values of maximum and minimum values of excessive object usage and excessive non-usage had been calculated, which is presented in Table 6.1.

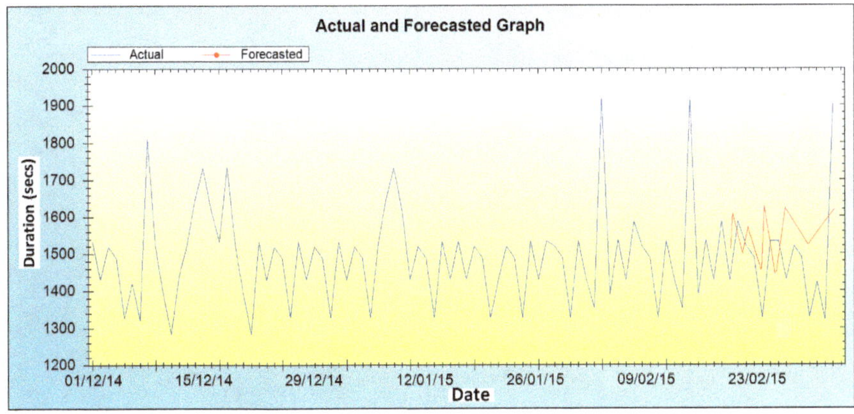

Fig. 6.6 Shower usage duration and forecasting for upcoming week

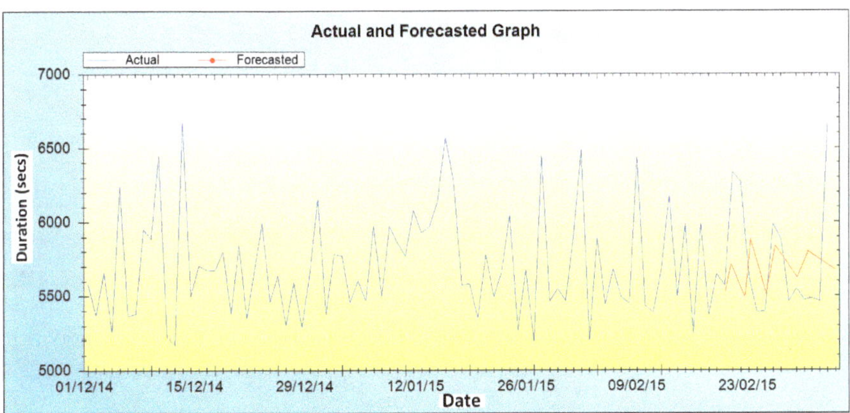

Fig. 6.7 Dining chair usage duration and forecasting for upcoming week

From the above table, it is identified that wellness forecasting system has forecasted three anomaly activities accurately. From subject three, there was two anomaly, one was related to bed over usage, and the other was related to a non-usage of a dining chair. That day the person slept longer; the occupant was not feeling well. Because of sickness, the occupant did not eat much and low occupancy of dining chair recorded. Subject four recorded anomaly for toilet over usage because of stomach upset of the occupant.

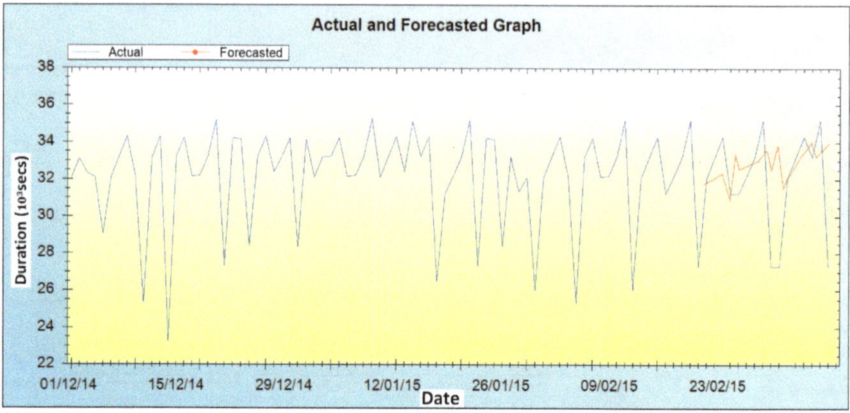

Fig. 6.8 Bed usage duration and forecasting for upcoming week

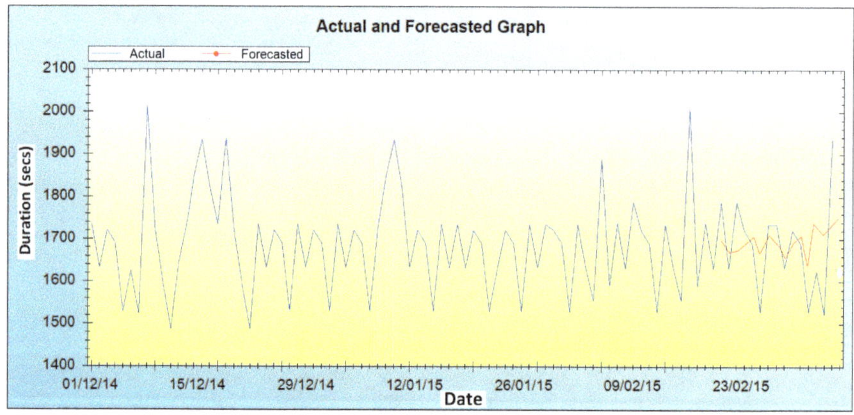

Fig. 6.9 Toilet usage duration and forecasting for upcoming week

6.3 Comparative Results

The majority of work related to an anomaly behavior pattern generation and detection is based on the identification of activity of daily living or near real time conditions. That indicates researchers use near real-time activity data to upper-level sensor context. Also, the use of machine learning and data mining models to detect the human behavior on the basis of hidden states is common among the researchers. One feature that distinguishes present wellness time series with existing research

Table 6.1 Wellness function indices of object usage and forecast of the ADLs

SUB	Activity	Sensor_ID	β_1	β_2	Forecasting for an upcoming week					Actual-duration (s)	Status
					Max-time (s)	Min-time (s)	α	δ	Γ		
1	Sleeping	Bed	0.77	0.884	29,483	23,483	0.200	0.120	0.421	24,317	Regular
	Dining	Chair		0.931	5580	3219	0.140	0.0583	0.490	3618	Regular
	Toilet	Toilet		0.882	1783	1247	0.0300	0.0580	0.700	1537	Regular
	Shower	S-Door		0.883	1501	1328	0.030	0.300	0.500	1433	Regular
	Relax	Couch		0.943	1658	1244	0.200	0.320	0.700	1329	Regular
	Watching TV	TV		0.921	3821	2750	0.059	0.200	0.300	3320	Regular
2	Sleeping	Bed	0.848	0.810	27,492	21,394	0.030	0.120	0.600	26,408	Regular
	Eating	Chair		0.860	4923	3028	0.200	0.070	0.400	3729	Regular
	Toilet	Toilet		0.823	1629	1249	0.048	0.380	0.550	1402	Regular
	Shower	S-Door		0.785	1420	1124	0.0200	0.400	0.350	1294	Regular
	Relax	Couch		0.842	1530	1204	0.100	0.540	0.530	1420	Regular
	Watching TV	TV		0.863	3502	2984	0.200	0.170	0.500	3320	Regular
3	Sleeping	Bed	0.884	**0.530**	**31,289**	**26,491**	**0.030**	**0.560**	**0.605**	**35,406**	**Anomaly**
	Eating	Chair		**0.510**	**4839**	**3429**	**0.300**	**0.460**	**0.170**	**1002**	**Anomaly**
	Toilet	Toilet		0.820	1509	1145	0.350	0.330	0.700	1329	Regular
	Shower	S-Door		0.889	1530	1239	0.120	0.300	0.500	1430	Regular
	Relax	Couch		0.943	1620	1307	0.100	0.459	0.300	1540	Regular
	Watching TV	TV		0.920	3309	2845	0.400	0.340	0.600	2984	Regular
4	Sleeping	Bed	0.792	0.883	29,042	25,302	0.150	0.300	0.400	27,404	Regular
	Eating	Chair		0.942	3942	2948	0.300	0.530	0.350	3720	Regular
	Toilet	Toilet		**0.550**	**1450**	**950**	**0.250**	**0.300**	**0.740**	**1730**	**Anomaly**
	Shower	S-Door		0.903	1730	1284	0.300	0.350	0.600	1630	Regular
	Relax	Couch		0.942	1430	1039	0.200	0.300	0.400	1240	Regular
	Watching TV	TV		0.840	3102	2503	0.400	0.300	0.600	2830	Regular

Table 6.2 Annotation used in confusion matrix table

S.no	Type of activity
A1	Bedtime
A2	Entrance
A3	Exit
A4	Water cattle
A5	TV
A6	Cooking (Electric oven, stove and grill)
A7	Eating-eating chair
A8	Microwave oven
A9	Fridge door
A10	Wash basin or kitchen hygiene
A11	Shower
A12	Toilet
A13	Computer use
A14	Sofa-relax

model is detecting activities in real-time with optimal accuracy. The present sensing data was collected from the urban home environment; the environment was affected by interference losses and error. There are a number of machine learning models that can be applied to present research objective, but three models were used to compare the quality of present research work (Lee et al. 2009; Eagle and Pentland 2009; Bakar et al. 2016).

These models are Hidden Markov models, conditional random fields (CRF) and naïve Bayes. Table 6.2 presents the annotation used in confusion matrix table.

Tables 6.3, 6.4, 6.5, 6.6, and 6.7 show the confusion matrix of different machine learning methods. Table 6.3 presents the confusion matrix for Naïve Bayes method; Table 6.4 presents confusion matrix for HMM and Table 6.5 presents the confusion matrix for CRF. Table 6.6 shows the confusion matrix for Wellness function. While Table 6.7 presents the comparative accuracy of all method included for comparison.

Table 6.3 Confusion matrix to show the accuracy of ADLs detection for different object usage by Naïve Bayes method

Actual	Predicted													
	A1	**A2**	**A3**	**A4**	**A5**	**A6**	**A7**	**A8**	**A9**	**A10**	**A11**	**A12**	**A13**	**A14**
A1	**55.03**	0	0	0	0	0	25.7	0	0	1.01	0.7	17.2	0	0
A2	0	**66.5**	33.01	0	0	0	0	0	0	0	0	0	0	0
A3	0	31.2	**68.62**	0	0	0	0	0	0	0	0	0	0	0
A4	0	0	0	**90.34**	0	0	0	0	0	0	0	0	0	0
A5	0	0	0	0	**90.1**	0	0	0	0	0	0	0	0	0
A6	0	0	0	15.2	0	**77.13**	0	7.3	0	0	0	0	0	0
A7	0	0	0	10.28	0	8.32	**44.12**	22.47	12.6	0	0	0	0	0
A8	0	0	0	0	0	0	0	**89.43**	0	0	0	0	0	0
A9	0	0	0	0	0	0	0	0	**88.53**	0	0	0	0	0
A10	0	0	0	0	0	0	0	0	0	**70.3**	0	0	0	0
A11	0	0	0	0	0	0	0	0	0	0	**66.38**	27.32	0	0
A12	0	0	0	0	0	0	0	0	0	0	40.28	**50.5**	0	0
A13	0	0	0	0	0	0	0	0	0	0	0	0	**95.31**	**0**
A14	0	0	0	10.28	0	9.32	24.47	0	10.6	0	0	0	0	**42.12**

Table 6.4 Confusion matrix to show the accuracy of ADLs detection for different object usage by HMM method

Actual	Predicted													
	A1	**A2**	**A3**	**A4**	**A5**	**A6**	**A7**	**A8**	**A9**	**A10**	**A11**	**A12**	**A13**	**A14**
A1	**61.53**	0	0	0	0	0	18.57	0	0	0.83	0.5	16.54	0	0
A2	0	**68.01**	29.45	0	0	0	0	0	0	0	0	0	0	0
A3	0	26.3	**69.2**	0	0	0	0	0	0	0	0	0	0	0
A4	0	0	0	**91.55**	0	0	0	0	0	0	0	0	0	0
A5	0	0	0	0	**91.18**	0	0	0	0	0	0	0	0	0
A6	0	0	0	10.68	0	**79.21**	0	6.09	0	0	0	0	0	0
A7	0	0	0	12.45	0	6.3	**45.3**	25.31	10.38	0	0	0	0	0
A8	0	0	0	0	0	0	0	**90.32**	0	0	0	0	0	0
A9	0	0	0	0	0	0	0	0	**90.32**	0	0	0	0	0
A10	0	0	0	0	0	0	0	0	0	**74.21**	0	0	0	0
A11	0	0	0	0	0	0	0	0	0	0	**65.74**	25.2	0	0
A12	0	0	0	0	0	0	0	0	0	0	37.4	**54.61**	0	0
A13	0	0	0	0	0	0	0	0	0	0	0	0	**97.32**	**0**
A14	0	0	0	12.08	0	9.58	20.17	0	8.6	0	0	0	0	**47.61**

Table 6.5 Confusion matrix to show the accuracy of ADLs detection for different object usage by CRF method

Actual	Predicted													
	A1	**A2**	**A3**	**A4**	**A5**	**A6**	**A7**	**A8**	**A9**	**A10**	**A11**	**A12**	**A13**	**A14**
A1	**60.61**	0	0	0	0	0	16.34	0	0	0.51	0.3	13.8	0	0
A2	0	**70.21**	27.9	0	0	0	0	0	0	0	0	0	0	0
A3	0	22.55	**73.28**	0	0	0	0	0	0	0	0	0	0	0
A4	0	0	0	**92.60**	0	0	0	0	0	0	0	0	0	0
A5	0	0	0	0	**92.41**	0	0	0	0	0	0	0	0	0
A6	0	0	0	7.38	0	**82.52**	0	5.81	0	0	0	0	0	0
A7	0	0	0	11.7	0	3.2	**47.7**	23.77	12.65	0	0	0	0	0
A8	0	0	0	0	0	0	0	**90.35**	0	0	0	0	0	0
A9	0	0	0	0	0	0	0	0	**92.87**	0	0	0	0	0
A10	0	0	0	0	0	0	0	0	0	**76.7**	0	0	0	0
A11	0	0	0	0	0	0	0	0	0	0	**64.27**	6.9	0	0
A12	0	0	0	0	0	0	0	0	0	0	35.54	**52.4**	0	0
A13	0	0	0	0	0	0	0	0	0	0	0	0	**96.78**	0
A14	0	0	0	8.034	0	12.50	19.15	0	9.08	0	0	0	0	**49.62**

Table 6.6 Confusion matrix to show the accuracy of ADLs detection for different object usage by Wellness method

Actual	Predicted													
	A1	**A2**	**A3**	**A4**	**A5**	**A6**	**A7**	**A8**	**A9**	**A10**	**A11**	**A12**	**A13**	**A14**
A1	**96.4**	0	0	0	0	0	0	0	0	0	0	0	0	0
A2	0	**97.73**	0.2	0	0	0	0	0	0	0	0	0	0	0
A3	0	0.51	**98.3**	0	0	0	0	0	0	0	0	0	0	0
A4	0	0	0	**98.99**	0	0	0	0	0	0	0	0	0	00
A5	0	0	0	0	**998.99**	0	0	0	0	0	0	0	0	0
A6	0	0	0	0	0	**998.99**	0	0	0	0	0	0	0	0
A7	0	0	0	0	0	0	**93.5**	0	0	0	0	0	0	0
A8	0	0	0	0	0	0	0	**97.99**	0	0	0	0	0	0
A9	0	0	0	0	0	0	0	0	**97.7**	0	0	0	0	0
A10	0	0	0	0	0	0	0	0	0	**96.4**	0	0	0	0
A11	0	0	0	0	0	0	0	0	0	1.3	**98.42**	0	0	0
A12	0	0	0	0	0	0	0	0	0	0	0.70	**97.57**	0	0
A13	0	0	0	0	0	0	0	0	0	0	0	0	**98.99**	**0**
A14	0	0	0	3.27	0	2.53	2.38	0	1.47	0	0	0	0	**92.28**

Table 6.7 Accuracy of different machine learning methods

Model	Accuracy
Naïve Bayes	75.09%
HMM	77.91%
CRF	79.55
Wellness	96.33%

6.4 Conclusion

This chapter presents the wellness based time series for wellness forecasting. The present chapter used the wellness formula derived in chapter five. Other existing models were able to forecast according to recent data but failed to include the dynamic change in behavior caused by seasons. The present chapter included the seasonal variations and offered the most appropriate curve fitting.

The recent data and historical data from last season have been applied to train wellness time series model. Wellness time series model measured the wellness of an occupant for upcoming days. Moreover, the comparison of wellness time series model with other most applicable machine learning algorithm was given for performance evaluation. Performance evaluation was based on the accuracy of activity recognition.

References

R. M. Pulsford, M. Cortina-Borja, C. Rich, F.-E. Kinnafick, C. Dezateux, and L. J. Griffiths, "Actigraph accelerometer-defined boundaries for sedentary behaviour and physical activity intensities in 7 year old children," *PloS one*, vol. 6, p. e21822, 2011.

J. L. C. Candás, V. Peláez, G. López, M. Á. Fernández, E. Álvarez, and G. Díaz, "An automatic data mining method to detect abnormal human behaviour using physical activity measurements," *Pervasive and Mobile Computing*, vol. 15, pp. 228–241, 2014.

M. T. Moutacalli, A. Bouzouane, and B. Bouchard, "The behavioral profiling based on times series forecasting for smart homes assistance," *Journal of Ambient Intelligence and Humanized Computing*, vol. 6, pp. 647–659, 2015.

U. Bakar, H. Ghayvat, F. Hasan, and S. Mukhopadhyay, "Activity and anomaly detection in smart home: A survey," *in Next Generation Sensors and Systems*, ed: Springer, 2016, pp. 191–220.

Ghayvat, H., Mukhopadhyay, S., Liu, J., and Gui, X.: 'Wellness Sensors Networks: A Proposal and Implementation for Smart Home to Assisted Living' *IEEE Sensors Journal,* 2015, Volume: 15, Issue: 12, pp. 7341 – 7348.

M.-S. Lee, J.-G. Lim, K.-R. Park, and D.-S. Kwon, "Unsupervised clustering for abnormality detection based on the tri-axial accelerometer," in *ICCAS-SICE, 2009*, 2009, pp. 134–137.

N. Eagle and A. S. Pentland, "Eigenbehaviors: Identifying structure in routine," *Behavioral Ecology and Sociobiology,* vol. 63, pp. 1057–1066, 2009.

Chapter 7
Conclusion and Future Works

Abstract The present chapter concludes and summarizes whole research, moreover it defines the possibility of extension in this research according to future demands.

This work shows the activity pattern extraction and prediction in a Wellness Protocol based AAL. The wellness environment was equipped with hydrogenous wireless sensors and networks. The research presented development of Wellness Protocol for smart home monitoring. The design and implementation of heterogeneous WSNs was described in detail. The designing of wireless intelligent sensing node, building the smart logic to overcome the issues and limitation of ZigBee was achieved. The proposed wellness function technique can forecast the next activity of the inhabitant, and give details of recent object usage. The Activities were extracted from the raw sensing events. Wireless sensors were deployed to record the behavior of the occupant and monitor the activities of daily living. The wellness pattern result generated in present research work was evaluated at different levels. The Wellness Protocol was developed as an integrated platform which enables software and hardware to monitor the ADLs of an occupant.

In general the major findings of the research work are listed below:

In most of the smart home the sensor data get affected by interference and losses present in home environment. There are number of existing smart home research but no one had ever considered the issue of interference by household appliances operating in the ISM band. Moreover the attenuation losses introduced by building material also degrades the sensing data. The present research discovered the significant source of interference and identified the possible mitigation solution. Deployment suggestions had been listed without demanding the change in building construction. Frequency spectrum analysis had been done to identify the suitable frequency channel for wellness system operation.

The analog sensing data received from the sensing systems attached to the various household appliances and other object was continuous and large. There was a need of intelligent transmission control algorithm to handle the large data. An intelligent sampling and transmission control algorithm designed for large data

H. Ghayvat and S.C. Mukhopadhyay, *Wellness Protocol for Smart Homes*,
Smart Sensors, Measurement and Instrumentation 24,
DOI 10.1007/978-3-319-52048-3_7

handling. The algorithm was based on event and priority based data transmission. This algorithm did not lose or block the useful data packets. The approaches presented in this thesis rely on the events data produced from the household appliances and object rather than continuous data. The system deployed and tested at different subject homes for assessing multiple occupant people behaviors who are residing at different homes.

The present research was designed and implemented a novel integrated framework for wellness detection of the individual. The wellness detection was done through generating the activity of daily living based on the interaction with different objects in a home environment. The activity detection, wellness pattern generation, and forecasting helped to identify any anomaly change in the lifestyle. The developed system was based on wellness protocol. The system had modified wellness indices β_1 and β_2 which reduced the false positive alarm at a significant level. The system was installed in four different houses; few of them were very old houses. The system was designed in such a way that the users did not need to look after its maintenance except small wear and tear, and power supply to sensing system.

Further investigation, in which future work could proceed, is listed below.

Smart homes are introducing advancement of embedded system and IOTs to monitor and control them, thus it is demanding a highly dynamic and complex system. Present system offered the monitoring but did not include the remote control over AAL up to full extent. This would also help to make home environment context aware.

A future research can be done to implement the proposed wellness system for multiple occupancy environments, where more than one inhabitant lives in one home. This is complex task to implement without using RFID.

The activity detection and forecasting techniques implemented in this work could be extended to detect and forecast more complex human behavior. For the aspects of future work, some other machine learning model would be investigated to list more complex human behavior.

The present research has only included fall detector sensor as body area network. There is still a need to introduce some other features of health care. Monitoring the human physiological parameters in real time can enhance the health safety.

Zeitfracht Medien GmbH
Ferdinand-Jühlke-Straße 7
99095 Erfurt, Deutschland
produktsicherheit@kolibri360.de